Populism and Postcolonialism

This book investigates the interconnections between populism and neoliberalism through the lens of postcolonialism. Its primary focus is to build a distinct understanding of the concept of populism as a political movement in the twenty-first century, interwoven with the lasting effects of colonialism.

This volume particularly aims to fill the gap in the current literature by establishing a clear-cut connection between populism and postcolonialism. It sees populism as a contemporary and collective political response to the international crisis of the nation-state's limited capacity to deal with the burst of global capitalism into everyday life. Writings on Ecuador, Colombia, Chile, Brazil, Italy, France and Argentina offer regional perspectives which, in turn, provide the reader with a deepened global view of the main features of the multiple and complex relations between postcoloniality and populism.

This book will be of interest to sociologists, anthropologists and political scientists as well as postgraduate students who are interested in the problem of populism in the days of postcolonialism.

Adrián Scribano is Director of the Centre for Sociological Research and Studies and Principal Researcher at the National Scientific and Technological Research Council, Argentina. He is also the Director of the *Latin American Journal of Studies on Bodies, Emotions and Society* and the Study Group on Sociology of Emotions and Bodies, in the Gino Germani Research Institute, Faculty of Social Sciences, University of Buenos Aires.

Maximiliano E. Korstanje is a global cultural theorist specialising in terrorism, mobilities and tourism. He is a Senior Researcher at the University of Palermo, Buenos Aires (Economics Department) and Editor in Chief of the *International Journal of Safety and Security in Tourism/Hospitality* (University of Palermo, Argentina).

Freddy Timmermann is a Professor at the Silva Henríquez Catholic University, Santiago de Chile; Bachelor in History and Professor of History and Geography (Pontifical Catholic University of Valparaíso); and Magister and Doctor in History (University of Chile).

Routledge Research on Decoloniality and New Postcolonialisms
Series Editor: Mark Jackson, Senior Lecturer in Postcolonial Geographies, School of Geographical Sciences, University of Bristol, UK.

This series provides a forum for innovative, critical research into the changing contexts, emerging potentials, and contemporary challenges ongoing within postcolonial studies. Postcolonial studies across the social sciences and humanities are in a period of transition and innovation. From environmental and ecological politics, to the development of new theoretical and methodological frameworks in posthumanisms, ontology, and relational ethics, to decolonizing efforts against expanding imperialisms, enclosures, and global violences against people and place, postcolonial studies are never more relevant and, at the same time, challenged. This series draws into focus emerging transdisciplinary conversations that engage key debates about how new postcolonial landscapes and new empirical and conceptual terrains are changing the legacies, scope, and responsibilities of decolonising critique.

Postcolonialism, Indigeneity and Struggles for Food Sovereignty
Alternative Food Networks in the Subaltern World
Edited by Marisa Wilson

Coloniality, Ontology, and the Question of the Posthuman
Edited by Mark Jackson

Unsettling Eurocentrism in the Westernized University
Edited by Julie Cupples and Ramón Grosfoguel

History, Imperialism, Critique
New Essays in World Literature
Edited by Asher Ghaffar

Arendt, Fanon, and Political Violence in Islam
Patrycja Sasnal

Populism and Postcolonialism
Edited by Adrián Scribano, Maximiliano E. Korstanje and Freddy Timmermann

For more information about this series, please visit: www.routledge.com/Routledge-Research-in-New-Postcolonialisms/book-series/RRNP

Populism and Postcolonialism

Edited by Adrián Scribano, Maximiliano E. Korstanje and Freddy Timmermann

Routledge
Taylor & Francis Group

LONDON AND NEW YORK

First published 2020
by Routledge
2 Park Square, Milton Park, Abingdon, Oxon OX14 4RN

and by Routledge
52 Vanderbilt Avenue, New York, NY 10017

Routledge is an imprint of the Taylor & Francis Group, an informa business

First issued in paperback 2021

British Library Cataloguing in Publication Data
A catalogue record for this book is available from the British Library

Library of Congress Cataloging-in-Publication Data
A catalog record has been requested for this book

ISBN: 978-0-367-18070-6 (hbk)
ISBN: 978-1-03-208764-1 (pbk)
ISBN: 978-0-429-05940-7 (ebk)

Typeset in Times New Roman
by Taylor & Francis Books

Contents

Illustrations

Figures

Tables

Contributors

Joanildo Burity is a Professorial Fellow on the Postgraduate Programmes in Sociology and Political Science, and Director of the School of Postgraduate Studies, at the Joaquim Nabuco Foundation, Brazil. He is also a Professorial Fellow at the Department of Social Sciences, Federal University of Pernambuco, Recife, Brazil. His research focuses on religion and politics, with a particular interest in its relationship to globalisation, culture and identity, public policy, associationism and social movements. Additionally, he has a keen interest in qualitative methodology and poststructuralist political theory. His current project studies transnational networks of social activism involving ecumenical groups in Brazil, Argentina and the UK.

Angélica De Sena is a Professor on the Sociology course at the Faculty of Social Sciences, UBA and the National University of La Matanza. She has a PhD in Social Sciences from the University of Buenos Aires. She specialises in the study of social policies and emotions, and the methodology of social research. She teaches undergraduate and postgraduate courses at different universities in Argentina and abroad. She is a researcher at the Gino Germani Research Institute (FCS-UBA) of the UNLaM, and coordinates the Study Group on Social Policies and Emotions (CIES). She is the director of the *Revista Latinoamericana de Metodología de la Investigación Social* (*ReLMIS*). She directs research projects at UNLaM and UBA. She has participated in various scientific meetings and has publications in scientific journals, book chapters and books in relation to social policies and methodology of social research.

Antimo L. Farro is full Professor of Sociology at the Department of Social Sciences and Economics, Sapienza University of Rome. He has published widely on social movements, urban sociology and migration studies. Among his most recent publications include the book *La città inquieta: Culture, rivolte e nuove socialità* (co-written with S. Maddanu) and articles in *Scuola Democratica* and *International Review of Sociology.*

Verònica Gisbert-Gracia is Adjunct Professor of Sociology at the University of Valencia. Her current areas of research include feminist collective actions, cultural studies and politics of affects. She is the author of several papers in leading scholarly journals as well as book chapters.

Maica Gugolati holds a PhD in Social Anthropology from the École des Hautes Études en Sciences Sociales (EHESS), Paris. Her academic interests are focused on the intersection between visual anthropology and performance studies. She animated an academic seminar about decolonial studies at the Maison des Sciences de l'Homme (MSH) in Paris, and has participated in several international conferences and residencies. She has recently published articles in the journals *African and Black Diaspora: An International Journal, Women, Gender, and Families of Color* and *Visual Ethnography,* as well as for Aica Caraibe du Sud (International Association of Critics of Art).

Lucía Herrera-Montero is an independent researcher based in Germany. She holds an MA in Literature and Cultural Studies from Universidad Andina Simón Bolívar (UASB) and a PhD in Latin American Literature from the University of Pittsburgh. She has been a faculty member at the Area of Communications of UASB (Maestría en Comuniciación de la Universidad Andina Simón Bolívar) and at UPS (Universidad Politécnica Salesiana, Quito), where she also served as Chair of the School of Communications and Director of the Area of Humanities. Her work explores the relationship between different sorts of contemporary Latin American narratives and marginal subjectivities which, from the standpoint of official and normalised discourses, are basically conceived as worthless and disposable bodies. Her ongoing research deals with the relationship between violence and marginal bodies in neoliberal contexts.

Luis Herrera-Montero is a Professor at the University of Cuenca, Ecuador. He has a BA in Applied Anthropology (Universidad Politécnica Salesiana, Quito), an MA in New Technologies Applied to Education (Autonomous University of Barcelona) and a PhD in Arts and Humanities (University of Jaen). He is a member of the Working Group Subjectivities, Critical Citizenships and Social Transformations, under the auspices of CLACSO. As well as teaching in the undergraduate and postgraduate programmes of universities in Ecuador, he has worked as a consultant for Ecuadorean public institutions as well as for the International Labour Organization.

Maximiliano E. Korstanje is a Senior Researcher in the Economics Department of the University of Palermo, Buenos Aires, and Editor in Chief of the *International Journal of Safety and Security in Tourism/Hospitality.* He is a global cultural theorist specialising in terrorism, mobilities and tourism. He has been Visiting Professor in the Centre for Ethnicity and Racism Studies (CERS) at the University of Leeds, TIDES at the University of Las Palmas de Gran Canarias and the University of La Habana. He works as an advisor and reviewer of different editorial projects for academic publishers such as Elsevier, Routledge, Palgrave Macmillan and Edward Elgar.

His latest book is *The Challenges of Democracy in the War on Terror* (Routledge).

Thomas Jeffrey Miley is Lecturer of Political Sociology in the Department of Sociology at the University of Cambridge. He received his BA from the University of California, Los Angeles (UCLA) and his PhD from Yale University. He has lectured at Yale University, Wesleyan University and Saint Louis University (Madrid); and he has been a Garcia-Pelayo Research Fellow at the Center for Political and Constitutional Studies in Madrid (2007–2009). His research interests include comparative nationalisms, the politics of migration, religion and politics and democratic theory. His research features in journals such as *European Journal of Political Research, European Politics and Society, Nationalism and Ethnic Politics* and *Nations and Nationalism*.

Alejandro Pizzi is Associate Professor of Sociology at the University of Valencia. His current areas of research include the sociology of work and labour relations, economic sociology and the sociology of social movements. He is the author of several papers in leading scholarly journals, books and book chapters.

Paola Rebughini is full Professor of Sociology of Culture at the Department of Social and Political Studies, University of Milan. She has published widely on social theory, cultural pluralism, youth and collective agency, including articles in the journals *Current Sociology* and *Social Science Information*.

Adrián Scribano is Director of the Centre for Sociological Research and Studies (CIES) and Principal Researcher at the National Scientific and Technological Research Council, Argentina. He is also the Director of the *Latin American Journal of Studies on Bodies, Emotions and Society* and the Study Group on Sociology of Emotions and Bodies, in the Gino Germani Research Institute, Faculty of Social Sciences, University of Buenos Aires. In addition, he serves as Coordinator of the 26 Working Group on Bodies and Emotions of the Latin American Association of Sociology (ALAS) and as Vice-President of the Thematic Group 08 Society and Emotions of the International Sociological Association (ISA).

Freddy Timmermann is a Professor at the Silva Henríquez Catholic University, Santiago de Chile. He has a BA in History from the Pontifical Catholic University of Valparaíso, and an MA and PhD in History from the University of Chile. He is a researcher specialising in the neoliberal process in the recent history of Chile. His books include: *The Pinochet Factor: Devices of Power, Legitimation, Elites: Chile, 1973–1980; Text Violence, Context Violence: Chile, 1973; The Great Terror: Fear, Emotion and Discourse: Chile, 1973–1980*; and *Neoliberalism in Multi-Disciplinary Perspective* (edited with Adrián Scribano and Maximiliano E. Korstanje).

Brett Troyan is a Professor of History at Cortland College, State University of New York. She obtained her doctorate in history from Cornell University. She has been a visiting scholar at the Latin American Studies Center at Cornell University and the Centre for Latin American Studies at Cambridge University, United Kingdom. She is the author of the book *Cauca's Indigenous Movement in Southwestern Colombia: Land, Violence, and Ethnic Identity in Colombia*. She has also published a number of peer-reviewed journal articles on the Colombian indigenous movement as well as essays on Afro-Colombians. She is currently working on a book manuscript on the civil war in Costa Rica.

Cécile Vermot is Assistant Professor in the cluster of biotechnologies in society (PBS) at Sup'Biotech PARIS (Engineering School), and Lecturer in the Université Des Patients at Sorbonne Université. She has a PhD in Sociology from the Université Paris Descartes and the Autonomous University of Barcelona. She is one of the coordinators of the International Network of Sociology of Sensibilities, with Adrián Scribano. Her research has focused on the interrelation between migration, emotions, gender and representations regarding the Argentinean migrants in Miami and Barcelona, leading to many articles published in French, English and Spanish. Recently, she edited a special issue on the thematic of the emotions of migrants at Migrations Société, and is currently writing a book about this subject.

Acknowledgements

We wish to thank all the contributors who have presented high-quality essays. Much thanks also to Faye Leerink and Routledge staff for their editorial assistance, and we are grateful to Majid Yar for his proofreading work. Those involuntary errors that remain in the text are ours. Our families deserve recognition for the time they generously donated while this project took place.

Introduction

Populism and postcoloniality: Geopolitical experiences

Adrián Scribano

The logic of capitalist reasoning expands and is metamorphosed across the entire planet under the multiple guises of populism. The central objective of this book is to shed light on the multiple and complex relations between populism and postcoloniality. From a plural and multidisciplinary perspective our investigation seeks to reflect on this problematic from the standpoint of diverse geopolitical and geocultural experiences.

The book enables an understanding of populism from a postcolonial perspective that makes it possible to: a) discuss the connections of populism as a political movement and the proximity and distances with the postcolonial demands of autonomy/dependence of people, collectives and nations; b) show some of the edges of the historical processes that lead us to the current situation of postcoloniality and emergence of "other" populisms; and c) provide a systematic perspective on the innumerable ways of understanding populism from a postcolonial perspective.

The different cultural backgrounds and origins of the authors ensures not only a high-quality product but also a multicultural viewpoint which helps us to understand this complex issue beyond the lens of bipolarity. In the book, views are gathered from political science, anthropology, history, and sociology, which enable the reader to gain a theoretical, epistemic and methodological vision of a complex problematic.

The phenomenon and the notion of populisms are elusive objects but both the oldest forms and those that we are discussing today across the entire world are related, among other things, to being processes that narrate themselves as an embodiment of the voice of the people.

Postcolonial studies are proposed, among other things, in order to break with the magic circle of the word of the dominator as the only tool to understand social processes.

The present book starts from the urgency of overcoming the countless misunderstandings and manipulations revolving around the term populism.

Populism: from history to concept

As is well known, populism is a slippery and very controversial term. The objective of this section is to set out successive approaches that allow us to understand the complexity of the phenomenon and capture some of its central features. As with all social phenomena, one should always remain alert to their changes in history and the theoretical transformations that refer to them. In this case, the history of populism is also the history of its conceptualisation.

Since the beginning of the social sciences the interest in understanding the regimes and/or political movements that express or cause systemic crises has been a topic of systematic enquiry. An example of this is Marx's study of Bonapartism and Weber's study of Caesarism. The numerous enquiries about National Socialism, fascism and its relations with the "state of exception" can be understood in a similar way.

Research and reflection on populism must address the connection between systemic crisis, political regime and changes in the political economy of morality. In this section I take a conceptual approach to populism and then articulate it with the contents and postcolonial developments.

Populism is a political, social and economic phenomenon that must be understood within the framework of the scenarios that appear. In a general way all forms of populism share some characteristics that qualify their contexts of appearance: a) the existence of an economic, political and/or social crisis; b) the emergence of "new" actors, classes and/or groups that dispute their participation in society and politics; and c) the presence of technological changes.

In broad strokes since in the last 150 years three great moments of populism have emerged: a) at the end of the nineteenth century, b) between the 1940s and 1960s; and c) the first years of the present century.

End of the nineteenth century

The two movements/parties that are taken as references to indicate the appearance of what we call populism today are Narodism in Russia and the "people's party" in the United States

The first was born in the internal tensions of Tsarism, its intellectual elites and bourgeoisie, as a movement that posits "going to the people" as a recovery of collective rural experiences. Lenin was at the same time an interpreter and participant in the disputes over the adequacy or otherwise of the populist explanations about capitalism and the ways of overcoming it. The essence of Narodism is that it represented the producers' interests from the standpoint of the small producer, the petty bourgeois. The "source" of Narodism lies in the predominance of the class of small producers in post-reform capitalist Russia:

> This description requires explanation. I use the expression "petty bourgeois" not in the ordinary, but in the political-economic sense. A small producer, operating under a system of commodity economy—these are the two features

of the concept "petty bourgeois," Kleinbürger, or what is the same thing, the Russian meshchanin. It thus includes both the peasant and the handicrafts-man, whom the Narodniks always placed on the same footing—and quite rightly, for they are both producers, they both work for the market, and differ only in the degree of development of commodity economy. Further, I make a distinction between the old and contemporary Narodism, on the grounds that the former was to some extent a well-knit doctrine evolved in a period when capitalism was still very feebly developed in Russia, when nothing of the petty-bourgeois character of peasant economy had yet been revealed, when the practical side of the doctrine was purely utopian, and when the Narodniks gave liberal "society" a wide berth and "went among the people." It is different now: Russia's capitalist path of development is no longer denied by anybody, and the break-up of the countryside is an undoubted fact. Of the Narodniks' well-knit doctrine, with its childish faith in the "village community," nothing but rags and tatters remain. From the practical aspect, utopia has been replaced by a quite un-utopian programme of petty-bourgeois "progressive" measures, and only pompous phrases remind us of the historical connection between these paltry compromises and the dreams of better and exceptional paths for the fatherland. In place of aloofness from liberal society we observe a touching intimacy with it. Now it is this change that compels us to distinguish between the ideology of the peasantry and the ideology of the petty bourgeoisie.

(Lenin, 1960, p. 395)

The second appears as a response of the peasants of the southern United States to the crisis of silver and grain prices that ended up being not only an "alter-native" to the Republican/Democrat binomial, but also a movement associated with the people. We can see here a critique of the political system but at the same time the selection of "concerns" and needs away from the people.

Following the ideas of Pierre Yared, it is possible to understand:

Historical research tends to view the late 19th century populist movement in the United States as caused by a wave of liberal ideas and long term social problems. However, many observers and activists of the time like William Jennings Bryan attributed discontent to economic conditions, particularly those of farmers, which they argued had been caused by the demonetization of silver. A simple analysis of agrarian price trends and political activity at the state level suggests that agrarian economic misery independent of monetary policy triggered the election of populist repre-sentatives who fought against the demonetization of the silver standard.

(Yared, 2003, p. 7)

Beyond the deep differences, two fundamental axes of populism emerge clearly: to speak on behalf of the people and to respond to a crisis in the system.

Between the 1940s and 1960s

From the 1940s to the 1970s, Latin America experienced many of the regimes that are considered "models" of populism. Explanations from the social sciences for the struggle of rising social classes, either as a result of modernisation processes or the changes in the political regimes associated with the Second World War, sought to understand the "originality" of these phenomena.

It was Gino Germani who conceptualised the national and popular movements as phenomena inscribed in the processes of modernisation, expansion of political participation and nationalist claims:

> This *sensation of participation* is not necessarily related to the effective influence that the popular classes can exert on the government, although, as we have indicated, the management has quite broad limits. Nor is there a close relationship between this sensation of participation and the economic improvements that these regimes are really capable of doing. In spite of the general opinion that the adhesion of the popular classes is obtained thanks to demagogic economic promises, the real foundation of popular support is the *"participation experience"*, which we have tried to describe.
>
> (Germani, 1973, p. 35; my translation and emphasis)

In the context of the processes of a revolution of expectations, increased consumption and dissolution of traditional rural patterns of authority, Germani's interpretation draws attention to the effective expansion of participation, the inadequacy of a demagogic characterisation, and the differences with fascism and National Socialism

For Torcuato Di Tella, populism is explained by the large-scale multiplication of the "incongruous" understood as a type of subject whose position and class condition has suffered a phase shift thanks to economic growth and modernisation:

> Populism, therefore, is a political movement with strong popular support, with the participation of non-working class sectors with important influence in the party, and sustaining an anti-status quo ideology. Their sources of force or "organizational links" are: I. An elite located in the middle or high levels of stratification and provided with anti-status quo motivations. II. A mobilized mass formed as a result of the *"revolution of aspirations"*, and III. An ideology or a widespread emotional state that promotes communication between leaders and followers and creates a *collective enthusiasm*.
>
> (Di Tella, 1973, pp. 47–48; my translation and emphasis)

From another perspective, Francisco Weffort recalls that populism occurs in a crisis of oligarchic hegemony that compels a commitment between different classes, among the bourgeoisie and the rising proletariat and others class fractions: "The peculiarity of populism comes from the fact that it emerges as a form of domination under conditions of 'political emptiness' in which no class has hegemony and exactly no class believes itself capable of assuming it" (Weffort 1968, p. 178; my translation).

The beginning of the twentieth century witnessed the agglomeration of multiple classes and interests that were gradually well orchestrated in the cause of the populism. Sensations, aspirations, sutures, emptiness and gaps are several paths that this stage of populism took that helps us to understand the origin of what will develop in the twentieth century as a politics of sensibilities.

The first years of the present century

With the rise of the so-called left populisms in Latin America in the first decade of this century and right-wing populisms in the second decade other phenomena and explanations have emerged.

To the multiple characteristics that have been indicated already we can add racism, dogmatism, homophobia, and so on as traits that in a different way are attributed to the "new" populisms. In this framework, different ways of approaching the phenomena and theoretical schemes to understand them have also emerged.

For Latin America, Carlos de la Torre rehearses a classification of populism that allows us to capture in broad strokes the changes that occur leading to the populism of the twenty-first century. After the conceptualisation of classical populism and neoliberal populism, he defines radical populism as follows:

> Radical populists of the twenty-first century are similar to classical populists in their politicization of social and economic exclusions. As in some classical populist experiences, political and social polarizations coincided. Similarly to neoliberal populists, they portrayed traditional political parties as the source of their country's ills, and contributed to the collapse of party systems. They linked neoliberal economic policies directly with liberal politics, practices, and values.
>
> (de la Torre, 2017, p. 210)

On the other hand, since the crisis of 2008 in Europe and the US, concern and interest in the various forms of populism have taken on new importance. In his paper "The Success of Extremist Parties in Europe" Michel Hastings maintains that:

> Populism is a political style which is a source for change based on the systematic use of rhetorical appeal to the people. In its discursive form, it is characterized by a programmatic minimalism but with a great symbolic plasticity which makes it a vector conducive to forge multiple and even heterogeneous indignations (ethno-cultural, anti-tax, anti-elitist, Eurosceptics, etc.). In its institutional form, populism includes partisan groups who intend to translate these empty statements into political regeneration projects by mobilizing the imagination of a virtuous people, who symbolizes the last repository of

national values, ready to follow a strong leader who embodies the axioms of transparency, proximity, similarity and truthfulness.

(Hastings, 2013, p. 9)

In this context, one of the most talked about visions for understanding the populism of the present century is that of Ernesto Laclau, who in the preface to his book *On Populist Reason* writes:

> One consequence of this intervention is that the referent of 'populism' becomes blurred, because many phenomena which were not traditionally considered populist come under that umbrella in our analysis. Here there is a potential criticism of my approach, to which I can only respond that the referent of 'populism' in social analysis has always been ambiguous and vague. A brief glance at the literature on populism—discussed in Chapter 1—suffices to show that it is full of references to the evanescence of the concept and the imprecision of its limits. My attempt has not been to find the true referent of populism, but to do the opposite: to show that populism has no referential unity because it is ascribed not to a delimitable phenomenon but to a social logic whose effects cut across many phenomena. Populism is, quite simply, a way of *constructing the political*.
>
> (Laclau, 2005, p. xi; my emphasis)

Empty signifiers, antagonism, demands, equivalences, hegemony and power are the characteristics of populism that in this way happens to be synonymous with the political. Beyond that, in this book reference is made to Laclau's theory in a more authoritative way than in this brief synthesis (see Chapter 1), and it is important to note that a modification of substances occurs from this perspective: populism is charged with positive (and "progressive") valences. Perhaps the identity populism = politics makes more understandable what is argued in Chapter 6 of this book regarding the connection between populism and neoliberalism.

These three historical moments of emergence and conceptualisation of populism provide us with a gateway to understand from a more conceptual perspective some of its main characteristics.

In the context described since the second half of the last century, the theoretical-analytical perspectives that propose to elaborate a postcolonial perspective on the phenomena of power and knowledge connected with the expansions of capitalism on a planetary scale have been consolidated. The next section tries to provide a synthesis of this.

Postcoloniality: a very short introduction

One of the strongest pieces of evidence for the presence of populism is the increasing presence of coloniality, and hence the urgency of retaking an analysis that can approach the phenomenon critically.

In the decade of the 1980s in English universities, a series of research units or academic departments were constituted around a multidisciplinary programme in social sciences that came to be called cultural studies. These groups included investigations and enquiries of diverse disciplinary origins. Among others, communication, literary criticism, sociology, anthropology and gender approaches can be mentioned. These collective efforts converged on the need to analyse the forms of domination in the central societies, crossed by difference and the inequalities of class, gender, ethnicity, religion and age.

Within the framework of the academic development of cultural studies and its multiplication in foreign universities, especially in the United States, research programmes with their own and sometimes independent characteristics were organised, such as multicultural approaches and subaltern studies.

Trying to synthesise the central characteristics of the different aspects of postcolonial theories, it is possible to affirm with Pajuelo that:

> Postcolonialism develops, thus, within the framework of the same conditions of possibility that discharged postmodernity, and in close relation to the profound "cultural turn" of the social and human sciences. Hence the reading of postcolonial discourse understood as an academic modality of postmodernism, and also its close relations with other currents of anti-hegemonic reflection such as "cultural studies", "subaltern studies" and "multiculturalism", of differentiated trajectories.
>
> (Pajuelo, 2001, p. 3)

One way to understand the current content of postcolonial options is to explore their "internal history", and one way to do it is to connect the interpretative schemes from those mentioned as "founding fathers" among which, and just to mention some, we must indicate Marcus Garvey, C. L. R James, Albert Memmi and Franz Fanon.

If the processes of institutionalisation and/or formalisation of groups/visions/approaches are taken as an indicator, two that should be mentioned are those of subaltern studies and the modernity/coloniality group (Quijano, 1988, 1993, 2000; Quijano and Wallerstein, 1992; Dussel, 1993, 1996, 2000; Mignolo, 1995, 2007; Lander; 2000, 2002; Escobar, 2007). The first visibly connects with the Guha proposal and the second includes Latin American authors who follow Guha's ideas (Guha, 1963; Guha, and Spivak, 2002).

On the other hand, the processes of reflection, conceptualisation and dialogue led to the reconfiguration of postcolonial studies, in decolonial approaches, and in perspectives from the standpoint of subalternity. The coexistence, proximity and distance of these viewpoints made the analysis more complex and limited the theoretical identities of each one (Coronil, 1996; Castro-Gómez and Mendieta, 1998).

It is at the beginning of the 2000s (with a deep and abiding connection with previous historical processes) that the different ways of approaching good living emerged. This revitalisation and redefinition added to the philosophical instruments and social theoretical tools the ancestral views about life (Santos, 2018).

Concomitant with the movement mentioned, feminist philosophy and theory complete, from a perspective of coloniality, racialisation and patriarchality, a set of perspectives "close" to postcolonial studies. It is not possible to address here properly this perspective, which has its own internal history.

In the tension of LGTBIQ, decolonial feminist and racialisation studies, we can also capture another space of proximity with postcolonial studies, as I want to show here (Lugones, 2002, 2003; Anzaldua, 2012; Tyagi, 2014).

In the aforementioned context, in parallel in some places and in others dependent on cultural studies, the so-called postcolonial studies were born and consolidated towards the end of the 1980s. The latter were presented towards the end of the 1990s as perhaps the most "radical" alternative to the canons of work in the social sciences. Defined from a multidisciplinary platform, where the narrative, historical and aesthetic-cultural is prevalent, postcolonial studies are tied around a group of simple common assumptions, but of great impact on the socially accepted discursive form of the social sciences. Among the most important assumptions we can mention: a) the criticism of European reason as the centre of scientific work; b) the possibility and the need to configure narratives from the "margins" crossed by class, gender and ethnic positions; c) the urgency to rethink particular realities from a geopolitics of knowledge; and d) the challenge of marking the terms of domination through the relationship between academia, science and everyday life.

The theoretical approaches that constitute postcolonialism can be defined in different ways. Pajuelo argues that these processes

> shaped the postcolonial reason as a series of theoretical practices based on the different colonial heritages, in the space of intersection between "local histories and global designs", a space in which decolonizing border epistemologies based on the local knowledge of the ex-colonial territories, in Europe, Africa, America, etc.
>
> (Pajuelo, 2001, p. 3)

It is possible to notice the following points in common between different visions: a) the relationship between particular history and knowledge; b) the connection between space and knowledge; c) the link between domination and knowledge; and d) the link between imperial culture, colonial culture and local knowledge.

In this way, a quick look at the contents of postcolonial studies allows us to understand that one of its clearest objectives is to demonstrate the ties of possible subjection between power and the scientific image of the world.

In this introductory chapter, I have been making an effort to understand the images of the world of analytical approaches to study Latin America (Scribano, 2004) and a post-independence perspective of construction of theories from the global South (Scribano, 2012). The starting point of these efforts can be summarised as follows.

The living force of capital, human beings turned mere "bodies-in-work" for the enjoyment of the few under the fantasy of a desire of all, needs to ensure the highest rate of ecological ownership in order to maintain in the medium term the (changing) structure of the ruling classes. In this regard, the location, management and treatment of water sources worldwide are among the edges of the predatory extraction strengthening its metamorphosis under conditions of inequality. The aforementioned inequality means no access to water supply or basic resources which grant the social reproduction; the biogenetic safeguards that are the necessary and sufficient conditions of the appropriation of the future. The consolidation of air and water extraction (in the context of processing, storage and distribution on a global scale) is based on the need for ownership of land that produces these two core elements of life. Jungles, forests and fields must be secured by the alliances of the fractions of the national ruling classes through guarantees of national states to private ownership, i.e. privatised and globalised international corporate environmental management.

In the same direction, the other edge of the extractive machinery is energy in all its forms, from oil to bodily energy, made socially consumable and available. Beyond the fatal process of extinction of these basic energies for capital, its regulation is currently the centre of reproduction in the short term. Therefore, a critique of eco-political economy is an important and indispensable step for understanding imperialist expansion. A constituent element of such a review is to make visible how bodily energies' policies meet, reveal and are written. The tribulations that numb bodies through social pain are one of the primary means to an unequal appropriation of the aforementioned bodily energies.

Second, for the current phase of imperialism, the production and handling of the regulation of expectations and the avoidance of social conflict are essential. Such management is ensured by the social bearability mechanisms and sensation control devices.

Third, imperial expansion involves in various ways the militarisation of the planet. There cannot be a balanced operation of extraction equipment and devices for the regulation of sensations without a repressive apparatus, a disciplinary and global control that transcends mere military occupation. This global repression aims to uphold the rule of neocolonial surveillance, given the paradoxical reorganisation of the compositions, class positions and conditions in complex space-times with centrifugal (away from the centre) and centripetal (towards the centre) movements of the various ways to resist the energy expropriation and regulation of feelings. In addition, the potential militarisation of all conflicts in geopolitically dependent systems is due to the metamorphosis of concentrated financial capital, the re-definition of corporative "patterns of accumulation", and both the fragmentation and unity of the expropriation.

In this way it may be understood, at least partially, how imperial expansion, characterised as an extractive apparatus of air, water, land and energy and as a repressive military machine, is sustained and reproduced, among other factors, by the production and handling of the regulation of sensations and mechanisms of social bearability.

In the same way, it can be understood how the linking texture between particularisations, individuation and locations are connected with the inequalities in the current state of the expansion of capital and imperialism, dependence and the colonial situation.

In the aforementioned context, questions, demands and criticisms about new forms of populism have emerged today. "Populism", "neopopulism", and "new rights" are some of the terms most used to characterise different political processes, parties and governing alliances in recent years across the planet. From the US through Italy and Europe to Latin America have emerged social, political and economic forces that are called or call themselves populism.

In the same sense, and also with a global character, it can be seen how the identification and/or stigmatisation of a social and political movement as populist has "legitimised" the rise of forces competing for the administration of the state "using" the populists as scapegoat for all evils.

This resurgence of populisms is inscribed in features of the contemporary social context: social networks, fake news and "non-truth". In this vein, the above noted landscape has led to a strange paroxysm which has created a re-symbolisation of what populism means. In a mixture of the old quasi-imperial manifestations of the past and the new "spontaneous" calls for social networks, these ways of doing politics are directly linked to the policies of the sensibilities in force across the planet.

Mass, people/*wolke*/*pueblo* and "homeland" are used as a platform for both defence and attack, are identified as a cause, but also as a solution to all problems. Imperialism, globalisation of markets and dependence are "reused" as explanatory concepts of the connections between political, social and economic systems.

The contents of the book

The central objective of this book is to shed light on the multiple and complex relations between populism and postcoloniality. From a plural and multidisciplinary perspective our investigation seeks to reflect on the problematic from the standpoint of diverse geopolitical and geocultural experiences.

The book begins with a chapter that reconstructs two axes of the current debate on populism: the political uses of the "people" and the presentation of the analytical proposal of Ernesto Laclau. Connecting religion, emotions and coloniality, the contribution of Joanildo Burity – "Populism, Religion and the Many Faces of Colonialism: Ongoing Struggles for 'the People'" – provides an opening to think of populism as a religious modality of understanding politics or a political modality of understanding religion.

Chapter 2 reconstructs two other important aspects of the populist phenomenon: its international character and the problematic of the "other" and the foreign. Cécile Vermot and Maica Gugolati, in "Parody, Satire and the Rise of Populism under Postcolonial Criticism: A French and an Italian Case", take up the rhetoric inscribed in the performance of satire and parody as vehicles to show the "sensibilities-in-game".

Maximiliano E. Korstanje writes the third chapter, titled "Neoliberalism and Populism in Argentina: Kirchnerism and Macrism as the Two Sides of the Same Coin", where he deals with two central axes of the current discussion on populisms: on the one hand asking about the existence or not of differences between left and right populisms, and on the other hand their distances and proximities with what traditionally is called neoliberalism. Returning to the recent Argentine experience, the author discusses the treatment of adversaries and antagonism to show the aforementioned axes.

Chapter 4, written by Alejandro Pizzi and Verònica Gisbert-Gracia, refers to Spain and is titled "Vox of Whom? An Approximation through Discourse Analysis and Study of the Profile of Its Social Base of Vox". From a quantitative perspective on discourse analysis, the essay reconstructs the narrative of a political party called Vox, addressing two fundamental topics in the discussion on European populism today: its fundamental narrative structure and those who share it. In the search for the sensitivities that are used, narrated and shared in writing, it provides a clarifying perspective on the narrative connection, emotions and "followers" of populism.

Brett Troyan elaborates Chapter 5 titled "From Jorge Eliécer Gaitán to Alvaro Uribe: A Brief Exploration of Populism in Colombia" where, from a historical perspective, he reconstructs the practices of the main Colombian political leaders and parties. Our author argues that:

> this chapter seeks to answer the question of whether there are continuities between this period of populism in the 1940s and with the conservative populist discourse of 2016 that led to the rejection of the peace agreement by the Colombian public. It argues that despite the very different political agendas some degree of continuity does exist in terms of political practices.

It is Angélica De Sena that in Chapter 6 addresses one of the most complex problems of all populism: social policies. In her contribution, "The Social Question in the Twenty-First Century: A Critique of the Coloniality of Social Policies", the author shows how on a planetary scale (with information from many countries in Asia, Europe, Africa, Latin America and even the United States) social policies are used to colonise desire and structure a politics of sensibilities. The success of colonial capitalism at the global level has in these state practices one of its fundamental bases to manage conflict.

In Chapter 7, called "Losing the Battle to Take Back Control? Clashing Conceptions of Democracy in the Debate about Brexit", Thomas Jeffrey Miley discusses one of the most important problems for Europe and the contemporary world: Brexit and democracy. The "failed" exit of the United Kingdom from the European Union shows how the "fear of the other" and the "claim of isolation" is transversal to populism and various forms of colonialism. As the author says: "[B]rexit is not only a manifestation of resurgent 'xeno-racism' among the populous, but simultaneously expresses and reflects a crisis of 'democratic leadership'."

Caudillism, paternalism, elitism and depoliticisation as features of a populism that emerges, is related to and connects with neoliberalism are the components of Chapter 8, entitled "Populism and neoliberalism in Chile", written by Freddy Timmermann, and which seeks to show some of the continuities between 1971 and the present in the complex Chilean reality. These are continuities of what the author calls neoliberal civilisation and evidence linked to the management of fear and the regulation of sensitivities.

In Chapter 9 written by Antimo L. Farro and Paola Rebughini, and titled "The Game of Disillusion: Social Movements and Populism in Italy", the authors characterise in a very accurate way the phenomenon that the book addresses from the Italian experience:

> Contemporary populism, well rooted in the digital media mechanisms, lives on contradictions, flattery, and immanence; it occurs in the absence of memory of what happened yesterday, and in the eschatological promise of a newness that is no longer emancipative, but just reassuring, again based on the promise that economic growth and social mobility will not stop. In this respect populist culture is a 'totality' unable to imagine the future, because its game is to play with the disillusions of the present.

In Chapter 10, Luis Herrera-Montero and Lucía Herrera-Montero in their "Intercultural Critical Reflections on Postcolonialist-Decolonialist and Populist Theories from Latin America and Ecuador", which is anchored in the Ecuadorian experience, suggest a postcolonial perspective that allows understanding an intercultural project that implies a decolonised consciousness and a new civilisational pact as the basis for a radical critique of populism. The text allows us to see in a different way the classic populist themes of clientelism and the fascination with the leader.

Finally, in my Chapter 11, called "Populism: The Highest Stage of Neoliberalism of the Twenty-First Century?", I propose as a goal "to make clear how the populism of recent years, through the globalisation of a regime of emotionalization, is possible to be understood as a highest stage of the global development of the neoliberalism". Through the notions of compensation and lumpen progressivism I propose to understand populism as the solutions to the capitalist crisis and not the other way around.

In the context of the above, it is possible to provide one of the many perspectives through which populism and postcolonialism are connected, trying to maintain a critical approach to critical politics.

At the beginning of his famous book *Fascism and Dictatorship* and in the context of his criticism of a phrase of Max Horkheimer, Nicos Poulantzas wrote, "he who does not wish to discuss *imperialism* … should stay silent on the subject of *fascism*" (Poulantzas, 1974, p. 17; emphasis in original). In this Introduction it is worth saying, "he who does not wish to discuss colonialism who should stay silent on the subject of populism". There is no capitalism without coloniality; populism being a capitalist phenomenon is a colonial practice.

Populism is colonial, beyond its rhetoric, for several reasons: it leaves intact the predatory exploitation structure of all forms of energy; it consecrates the fantasy of international autonomy basing the financing of the consumption that it makes in the sale of commodities and the consolidation of financial capital; and it develops a politics of sensibilities based on the management of emotions that narrate a normalised society in the immediate enjoyment through consumption.

Both in the model of the so-called left or right populisms the depredation of ecological assets, the dispossession of vital energy sources, the expropriation of nutrients and the almost-total private control of living beings and non-human individuals remain consecrated, naturalised and immune to criticism.

Populism today promotes mega mining, the extraction of coltan and so-called blood minerals, the exploitation of lithium, the shale oil and gas business among many other activities that involve the colonial presence of companies from central countries in poor or emerging regions and countries.

On the other hand, the parties or movements identified with populism do not intend to transform or modify the appropriation of fresh water by a very small group of private corporations across the world, nor the expansion of monocultures such as soy, nor a serious criticism of global warming in terms of a geopolitics of emissions and pollution, and there is no clear position regarding wildlife and natural forests and their devastation by commercial interests.

In the same vein, nothing is said about private patents on bio-diversity, the concentration of production, management and distribution of food, the geopolitical appropriation of seeds, lands and climates that provide macro and micronutrients, as well as the modification of plant and animal genetics and their concentration in a few international corporations.

The colonial depredation accepted and naturalised by populism as capitalist resignation leaves unchanged the appropriation of all forms of life through genetic design, nanotechnological modifications and the management of sensibilities in and through many of the features of the technology 4.0 revolution

It is in this context that the acceptance by the populists of the existence of a normalised society in the immediate enjoyment through consumption consecrates and reproduces the coloniality of everyday life. It is at this point that policies to promote consumption, strategies to give access to consumption to an increasing number of people, and programmes aimed at the management of socio-emotional skills as a bastion of new forms of adaptability and flexibility match populism and colonialism.

It would be a political and analytical error to think that the reality that we finish delineating has the character of a closed totality, of immanence substantially unmodifiable, of a moral political economy without breaks. As I have pointed out in other places (Scribano, 2015, 2017), there is a set of interstitial practices that tear and break the pretension of totality of systemic reason: filial love, collective trust and reciprocity.

References

Anzaldua, G. (2012) *1987 Borderlands/La Frontera: The New Mestiza.* San Francisco: Aunt Lute Books.

Castro-Gómez, S. (1996) *Crítica de la razón latinoamericana*. Barcelona: Puvill Libros.

Castro-Gómez, S. (ed.) (2002) *La reestructuración de las ciencias sociales en América Latina*. Bogotá: Universidad Javeriana.

Castro-Gómez, S. and Mendieta, E. (eds) (1998) *Teorías sin disciplina, latinoamericanismo, poscolonialidad y globalización en debate*. Mexico City: Miguel Angel Porrúa/University of San Francisco.

Coronil, F. (1996) 'Beyond Occidentalism: Toward Nonimperial Geohistorical Categories', *Cultural Anthropology*, 11(1), pp. 51–87.

de la Torre, C. (2017) 'Populism in Latin America', in C. Rovira Kaltwasser, P. Taggart, P. Ochoa Espejo and P. Ostiguy (eds) *The Oxford Handbook of Populism*. Oxford: Oxford University Press, pp. 79–100.

Di Tella, T. (1973) 'Populismo y reformismo', in G. Germani, T.S. Di Tella and O. Ianni (eds) *Populismo y contradicciones de clase en Latinoamérica*. Mexico City: Ediciones Era, pp. 38–83.

Dussel, E. (1993) 'Eurocentrism and Modernity', in J. Beverly and J. Oviedo (eds) *The Postmodernism Debate in Latin America*. Durham: Duke University Press, pp. 65–76.

Dussel, E. (1996) *The Underside of Modernity*. Atlantic Highlands, NJ: Humanities Press.

Dussel, E. (2000) 'Europe, Modernity, and Eurocentrism', *Nepantla*, 1(3), pp. 465–478.

Escobar, A. (2007) 'Worlds and Knowledges Otherwise', *Cultural Studies*, 21(2–3), pp. 179–210. doi:10.1080/09502380601162506

Germani, G. (1973) 'Democracia representativa y clases populares', in G. Germani, T. S. Di Tella and O. Ianni (eds) *Populismo y contradicciones de clase en Latinoamérica*. Mexico City: Ediciones Era, pp. 12–37.

Guha, R. (1963) *A Rule of Property for Bengal: An Essay on the Idea of Permanent Settlement*. Paris: Mouton & Co.

Guha, R. and Spivak, G. C. (2002) *Subaltern Studies Modernità e (Post)colonialism*. Verona: Ombre Corte.

Hastings, M. (2013) 'The Success of Extremist Parties in Europe', in I. Durant and D. Cohn-Bendit (eds) *The Rise of Populism and Extremist Parties in Europe*. The Spinelli Group. www.spinelligroup.eu/sites/spinelli/files/finalpopulismen_0.pdf

Laclau, E. (2005) *On Populist Reason*. London: Verso.

Lander, E. (ed.) (2000) *La colonialidad del saber: eurocentrismo y ciencias sociales*. Buenos Aires: CLACSO.

Lander, E. (2002) 'Los derechos de propiedad intelectual en la geopolítica del saber de la sociedad global', in C. Walsh, F. Schiwy and S. Castro-Gómez (eds) *Interdisciplinar las ciencias sociales*. Quito: Universidad Andina/Abya Yala, pp. 73–102.

Lenin, V. I. (1960) 'A Criticism of Narodnik Sociology: The Economic Content of Narodism and the Criticism of it in Mr. Struve's Book (The Reflection of Marxism in Bourgeois Literature)', in *Lenin Collected Works*, Volume 1. Moscow: Progress Publishers, pp. 333–508.

Lugones, M. (2002) 'Impure Communities', in P. Anderson (ed.) *Diversity and Community: An Interdisciplinary Reader*. Oxford: Blackwell, pp. 58–64

Lugones, M. (2003) *Pilgrimages = Peregrinajes: Theorizing Coalition Against Multiple Oppressions*. Lanham: Rowman & Littlefield.

Mignolo, W. (1995) *The Darker Side of the Renaissance: Literacy, Territoriality and Colonization*. Ann Arbor: University of Michigan Press.

Mignolo, W. D. (2007) 'Introduction', *Cultural Studies*, 21(2–3), pp. 155–167. doi:10.1080/09502380601162498

Pajuelo, T. R. (2001) 'Del "Postcolonialismo al Posoccidentalismo: una lectura desde la historicidad Latinoamericana y Andina"', *Comentario Internacional: revista del Centro Andino de Estudios Internacionales*, 2, pp. 113–131.

Poulantzas, N. (1974) *Fascism and Dictatorship*. London: Verso

Quijano, A. (1988) *Modernidad, Identidad y Utopía en América Latina*. Lima: Sociedad y Política Ediciones.

Quijano, A. (1993) 'Modernity, Identity, and Utopia in Latin America', in J. Beverly and L. Oviedo (eds) *The Postmodernism Debate in Latin America*. Durham: Duke University Press, pp. 140–155.

Quijano, A. (2000) 'Coloniality of Power, Eurocentrism, and Latin America', *Nepantla: Views from South*, 1(3), pp. 533–580.

Quijano, A. and Wallerstein, I. (1992) 'Americanity as a Concept, or the Americas in the Modern World-System', *International Social Science Journal*, 134, pp. 459–559.

Santos, B. de S. (2018) *The End of the Cognitive Empire: The Coming of Age of Epistemologies of the South*. Durham: Duke University Press.

Scribano, A. (2004) *Combatiendo Fantasma: Teoría Social Latinoamericana, una Visión desde la Historia, la Sociología y la Filosofía de la Ciencia*. Santiado de Chile: Universidad de Chile, Facultad de Ciencias Sociales.

Scribano, A. (2010) '*Las Prácticas del Querer: el amor como plataforma de la esperanza colectiva*', in M. Camarena and C. Gilabert (eds) *Amor y Poder: Replanteamientos esenciales de la época actual*. Mexico: Razón y Acción, pp. 17–33.

Scribano, A. (2012) *Teorías sociales del Sur: Una mirada post-independentista*. Buenos Aires: ESEditora.

Scribano, A. (2015) 'La Esperanza como contracara de la Depredación: Notas para una defensa del futuro', *Revista Actuel Marx/Intervenciones*, 19, pp. 175–193.

Scribano, A. (2017) 'Amor y acción colectiva: una mirada desde las prácticas intersticiales en Argentina', *Aposta. Revista de Ciencias Sociales*, 74, pp. 241–280.

Tyagi, R. (2014) 'Understanding Postcolonial Feminism in Relation with Postcolonial and Feminist Theories', *International Journal of Language and Linguistics*, 1(2), pp. 2374–8869.

Weffort, F. (1968) *O populismo na política brasileira*. Rio de Janeiro: Paz e Terra.

Yared, P. (2003) 'The Political Economy of Agrarian Participation in the Populist Movement'. Online at: www.semanticscholar.org/paper/The-Political-Economy-of-Agrarian-Participation-in-Yared/29f51bae38ea946e3ad0440e612ef443b905fbd7 [Accessed 26 April 2019]

1 Populism, religion and the many faces of colonialism

Ongoing struggles for "the people"

Joanildo Burity

A struggle for "the people" is under way across Latin America, and religion is very much part of it, whether through emerging actors or the underlying logics of the dispute. For nearly three decades a vigorous, but non-linear, process of pluralisation has made way for emerging minority social actors. Among the latter, *evangélicos* have featured permanently and controversially. Having grown significantly (though at different rates in the region), this religious minority rose to public visibility bent on redrawing cultural and political boundaries and facing head-on their identified adversaries. One way of understanding this phenomenon is to look at it as a double self-inscription: onto the national identity/culture (as opposed to being seen as an alien religion) and onto the public arena of debate, leadership and government.

Such a process has been predictably shot through with controversy and contestation but has fairly clearly succeeded. *Evangélicos* – and particularly, among them, as a majority within a minority, Pentecostals – are now both recognised and reckoned with. Emerging at the juncture of a general expansion of claims to representation and participation, following the return to civilian rule and democratisation of Latin American societies, they are in the game of constructing "the people". This raises questions about the recent politicisation of religion in Latin America in connection with what Laclau called "the populist reason". How to understand evangelical/Pentecostal discourse and public engagements? To what extent do this clearly popular form of religion and its ordinary mouthpieces and activists express an undoing of subalternity? How can one understand the clear divide between them and already-known activist forms of religious practice and the disputes between the two camps? What sort of politicisation is this, in the face of recent developments in which evangelicals/Pentecostals seem to represent the hard core of an extreme right politics in Brazil and not unrelated positions in other Latin American countries?

In what follows, I will explore these issues as demonstrated in three related cases: the emergence of Pentecostalism as popular and public religion; ecumenical transnational networks of socio-political activism; and the travails of democratisation in the context of a new religion/capitalism assemblage in the South akin to what William Connolly called the "evangelical-capitalist resonance machine" (Connolly, 2008). Given the salience of recent epistemic debates on the situatedness and critical edge of knowledge, which search for an authoritative voice from the Latin American context, I will begin with some theoretical notes on the debate on postcoloniality in relation to social and political activism of minorities and religion. I will then explore the two forms of self-inscription mentioned above and move on to examine their connections to the clear configuration of a fierce religious right-wing discourse at odds with older forms of Protestant public engagement in the region that could be called ecumenical.

Subalternity, decolonisation, populism and the recent rise of minority religion in Brazil

The contemporary debate on subalternity and decolonisation in Latin America has nothing to do with the mere assertion of the legitimacy of a "deficient" path in relation to a metanarrative of North Atlantic historical development – liberal, capitalist and Eurocentric, some sort of postcolonial pride. Rather, it is about a double movement: (a) the refusal of the parameters of analysis and evaluation that articulate this metanarrative; and (b) the identification of the same processes previously qualified as "deficient" in the societies that would be loci of this metanarrative.

Thus, on the one hand, the subalternising narrative is subtracted, returning it as criticism and diagnosis of the societies that produced it. An agency is asserted that "levels" the field by articulating a voice and assigning its supposedly deficient features to the majority order. The (re)emergence of populism, notably right-wing in the last fifteen years in Europe and, in recent years, in the United States, reveals a symptom of this North Atlantic "deficiency". There are others, such as the self-assertion of ethnic and religious minorities in the heart of these societies, who are increasingly impossible to identify as merely "external" to them (migrants, terrorists, missionaries), whether for their settlement (diasporas comprising several generations, national or religious identities that have long been present and now reactivated or transfigured), or for their wish to remain and to claim legal, political and cultural hospitality.

The critical edge of the decolonisation argument today lies in the understanding of the local, cross-border and global dynamics historically experienced by peripheral, postcolonial societies as a plot that spreads and traverses so-called central societies. As in the colonial past itself, this occurs despite widespread epistemic triumphalism and cultural and political arrogance in the metropolises and their academies. The experience of prolonged crises, never resolved by incomplete transitions, or of disaggregative ruptures, bellicose and/or technocratic political authoritarianisms, so often associated with the postcolonial world, became an inseparable part of the old colonising societies. The puncturing of its physical and symbolic boundaries by instability; multitemporal practices and ways of life; identity heterogeneity; the need for engagement in difficult and uncertain negotiations to (re)conquer basic rights or prevent them from being subtracted; difficulty in resisting processes that "come from without" and impose themselves in a naturalised way; and reversibility of achievements and stabilisations are all marks of our time. They thus allow for the "return" of the colonial gaze onto the colonisers themselves. But they also constitute conditions and procedures by means of which a subordinate voice is affirmed in search of autonomy, justice and recognition. Places and agencies traditionally seen as adversaries or indexes of an outdated past are once again activated, supporting the articulation of that voice, or rather, a plurality of voices.

It is not possible to reproduce here the details of this operation, which is at once epistemic and sociopolitical. I want to focus on two aspects in this work: the discourse on the "deficient" nature of postcolonial politics – exemplified in the signifier "populism" – and the construction of the imagination and social bond – exemplified in the signifier "religion". In the following paragraphs of this section, I will try to articulate the decolonising gesture concerning these two signifiers, which, given the simultaneity of the critical, epistemic and sociopolitical operations, must be seen not only as *concepts* but also as *signifiers* rhetorically managed by different social actors. If this section will focus more on the conceptual articulation of the two terms, the following will explore its interface with the performative and symbolic articulation.

Let us begin with a succinct presentation of the question of populism. The academic *doxa* is blatantly critical of what the term describes: populism is a heteroclite set of political practices of mobilisation and manipulation of the masses or a multiple and contradictory ideology, in which the defence of the people is often mounted by those who are not even part of it, as regards existing social hierarchies. Populism is demagoguery, mobilisation of passions, disrespect for institutions, and legal and representational procedures built in the wake of liberalism and socialism. Populism would be indicative of backward stages of political development, pre- or undemocratic. In the last decades, populism has also become a term of accusation against counter-cyclical economic policies, redistributive social policies, cultural policies of recognition, and policies of expansion and stimulation of social participation. It slowly grew into the big enemy of neoliberal and institutionalist discourses.

The discourse on populism, anyhow, clearly expresses an academic (and liberal political opponents') uneasiness about the affirmation or defence of "the people" articulated by those who claim to be populist or are accused of being so. It is thus inseparable, in its intellectual as well as political enunciations, from the determination of this contested entity, the people. Contrary to this interpretation, the work of Ernesto Laclau and a number of others referenced in it or in (critical) dialogue with it (Laclau, 2005; Barros, 2009; Arditi, 2014; Laclau, 2014; Biglieri, 2017; De Cleen et al., 2018; Finchelstein and Urbinati, 2018), insists on a *formal* approach to populism: instead of distilling a common meaning or conceptual core present in all/most cases identified as populist, or focusing on it as a movement, analysis should focus on the formal traits of the phenomenon. Populism is a "political logic" (Laclau, 2005, p. 150).

The language used by Laclau can raise questions. First, given the oscillation between "*a* people" and "*the* people". Second, because of the use of people between scare quotes. This could be seen as an ambivalence to be corrected. But as it happens, the second aspect denotes only the maximum degree of extension of the chain of equivalences that has already begun in the articulation of something less than a universal representation of the reconciled or emancipated community, or of a project of radical transformation of society. The distinction can actually be productive and highlight distinct instances of application. Let me start by quoting some of Laclau's formulations:

[a] when relations of equivalence between a plurality of demands go beyond a certain point, we have broad mobilisations against the institutional order as a whole. We have here the emergence of the 'people' as a more universal historical actor, whose aims will necessarily crystallise around empty signifiers as objects of political identification. There is a radicalisation of claims which can lead to a revolutionary reshaping of the entire institutional order.

(Laclau, 2014, p. 151)

[b] The *sans papiers* want to have *papiers*, and if the latter are conceded by the state, they could become one more difference within an expanded state. In order to become 'universal', something else is needed – namely, that their situation as 'outsiders' becomes a symbol to other outsiders or marginals within society; in other words, that a contingent aggregation of heterogeneous elements takes place. This aggregation is what I have called a 'people'.

(Ibid., p. 159)

[c] My argument is that the construction of the 'people' as a collective actor entails extending the notion of 'populism' to many movements and phenomena that traditionally were not considered so. And, from this viewpoint, there is no doubt that the American civil rights movement extended equivalential logics in a variety of new directions and made possible the incorporation of previously excluded underdogs into the public sphere.

(Ibid., pp. 173–174)

The distinction between "the people" and "a people", in quotes *a* and *b* above, the use of the signifier "people" in quotation marks and the attribution of the concept of populism (here in quotation marks, in unusual fashion in Laclau's work) warrant two conclusions that will be consequential in my subsequent discussion of the question of religion: (a) if "the people" is the name of a politically constructed collective actor, this actor should not be confused with the mass of citizens of a national state or with its general population, and must be specified, characterised, narrated contextually; (b) "a people" describes an equally politically constructed collective actor, but one that has not asserted itself or become the locus and the name of a *general* refusal of the institutional order, presenting itself as *aspiring* to this function, but still situated on a terrain where *other actors* may be more successful in this intent. More than that, "the people", Laclau argues in *Populist Reason*, strictly speaking, is not a referent of a group, but a way of building the unity of the group. It is not an ideological expression, but a form of relationship between social agents (Laclau, 2005, p. 97). How, then, does this construction take place?

Laclau does not start with a substantive definition of the people, but proposes that, as we are faced with a relational process, the construction of the people, of a people, be analysed from the articulation of demands, or rather from the passage of requests to claims, due to non-attendance or dissatisfaction with the form or the extent of fulfilment of these demands by the institutional order. The fulfilment of demands includes them as differences internal to the system, thus "pacifying" them. Rejection or partial compliance with them may lead, depending on the rhythm, delay or virulence with which the demands are answered, to the perception that there are other demands equally unsatisfied, partially met in an unjustified or unacceptable way, or plainly dismissed, and that they are equivalent through their common opposition to the established order. It is the *crystallisation* of these equivalences and their opposition to a common enemy that Laclau identifies as the elementary process of populism's formation.

We will see in the last section that the rising tide of recent right-wing populisms in Europe, the United States and Brazil indicate a variant of this formal description, which refers to the manifestation of *reactive responses* to experiences of identity displacement of traditional actors or failure of the hegemonic order. But here, it is worth outlining the basic contours of the Laclauian discursive approach. It is the equivalential articulation of unmet needs, constituting "a broader social subjectivity" (Laclau, 2005, p. 99), which defines the beginning of the constitution of a people "as a potential historical actor" (ibid.). But only the beginning. This is not yet a sufficient characterisation. For the articulation is not automatic or independent of an agency and the emergence of a "stable system of signification" (ibid.). A project that points to and seeks to institute a new hegemony: a "popular identity" (ibid., p. 102).

Distinct from an institutionalist articulation, which "attempts to make the limits of discursive formation coincide with the limits of the community" (ibid., p. 107), as differences within the system, populist articulation draws a boundary dividing society into two camps. "The 'people' in this case is something less than the totality of the members of the community: it is a partial component which nevertheless aspires to be conceived as the only legitimate totality" (ibid., pp. 107–108). In populism, a party that sees itself as underprivileged, a *plebs*, intends to be/represent all the members of the community, the *populus*, in the terminology inherited from Roman law: a part for the whole (ibid., p. 108).

I have argued, in previous works (Burity, 2015a, 2015b, 2016, 2017), that in the context of a broad activation of political subjectivities alongside the democratising processes across Latin America, since the 1980s, there has been a growing politicisation of religion. This has brought to light new public actors, the most successful being the *evangélicos* (largely comprising Pentecostals, but also naming historical Protestants). Brazil has been the most significant case in the region, but the examples are everywhere (Freston, 2008; Barrera Rivera and Pérez, 2013; López Rodriguez, 2014; Wynarczyk et al., 2016; Pérez Guadalupe, 2017). Given the invisibility of Pentecostalism as a public phenomenon, and the internal competition with traditional Protestantism, the activation of a politicised Pentecostal identity and what came after involved a process of self-assertion that was from the beginning both relational and contested. It was a response to the possibilities opened by the final stages of the struggle for democracy as of the late 1970s, but also the expression of their own fear of a left-wing advance that might restrict religious freedom and of mistrust from politicians and radical social activists who would only see in them conservatism, subservience and a likely threat of "confessionalisation of politics".

In carving a space for themselves in the midst of myriad movements, initiatives and group dynamics claiming a home in the rising democratic order, evangelicals argued for their own recognition as legitimate players in the game, called for equal treatment of all religions (in effect minoritising the Catholic Church, to count politically as one religious alternative among others) and contributed towards certain progressive outcomes of constitutional design. They also closed ranks against Afro-Brazilian religions as demonic practices.

Many have stressed the flexibility and plasticity of Pentecostalism as a religious alternative. Its weak confessional basis, stressing personal experience and ritual involvement over doctrinal formulation, would help understand its adaptive potential in highly adverse situations. It also favoured pragmatism and institutional simplicity, making it more portable (Corten and Marshall-Fratani, 2001; Anderson, 2004; Bergunder, 2009; Yong, 2010; Dempster et al., 2011; Medina and Alfaro, 2015). Although this profile changed dramatically over time, Pentecostalism/Evangelicalism, as a minoritised religion, started with claims to recognition and inclusion, participation and influence. In that respect, it was part of the newly emerging post-dictatorship people. Its trajectory shows multiple articulations of faith and public engagement, right and left, from ideological, activist or party-alignment perspectives, while keeping a hard core of moral and "theological" conservatism (Biblicism, proselytism and a rigorous personal ethos). Through such a moral conservatism as well as strong community ties, Pentecostals sought to achieve trust and goodwill from existing political and intellectual elites and the wider population, as they worked by example to provide evidence of an inchoate social renewal proposition.

As will be seen later, this intent quickly proved too fragile and inconsistent to withstand the pressures of political life and public controversies. Though deeply rooted in the story of Latin American Pentecostalism, such communitarian spirit and strict personal uprightness could not hold against the sway of prosperity gospel theology, internally, and entrepreneurial, business-oriented and consumerist discourses, externally (Burity, 2017; 2018a). This ushered in a new stage in Pentecostal public profile, which will be discussed in the last section.

A lot less visible than Pentecostalism grew to be, *ecumenical Protestant networks and organisations* represent a sharply distinctive mode of engagement. Rooted in mid- to late-twentieth-century ecumenical traditions of theological reflection on international development, radical politics and national liberation of former colonies in Africa and Asia, ecumenical social activism portrays a very different understanding of the people. A lot closer to left-wing constructions of grassroots mobilisation and liberation theology-inspired ecclesiological innovations, this form of religious public engagement parallels social-movement repertoires of action and tends to merge with other forces within this field, making it hard to spot with a naked eye. It also incorporates and translates back into the churches the identities and agendas of contemporary minority movements, particularly ethnic, gender-based and ecological ones. Strictly minoritarian in shape and approach, ecumenical social activism seems to take very seriously the gospel images of yeast in the dough or the seed that must die so that a new plant can sprout to life. It rarely makes any claim to protagonism or resorts to openly religious public language, though both can be found in various contexts, when local actors are too inorganically organised to voice their views and claims loud and clear. Given its self-ascribed reconciling mission, ecumenical social activism sometimes hangs on an uneasy balance between pluralistic accommodation and antagonistic denunciation of injustices, authoritarianism and violation of human rights. However, it does not refrain from exercising a "prophetic voice" and sharing in organisation and conduction of gatherings, protests and high-level political mediation, using strategically its international connections and global structures.

Both forms of Protestant public presence, therefore, promote a double self-inscription in articulating their understanding of the people: they lay a claim on national identity/culture (as opposed to being seen as alien forms of religion) and on the public arena of debate, leadership and government. I will say a bit more on these in the next two sections.

Evangélicos as part of the national people

Populism has retained a persistent association with negative views about the people, how it is formed, who leads or represents it, and how it relates to representations of the community or the sovereign political order (national societies and states, in modern terms). These negative connotations are only matched by the equally negative views of the nature, roles and positions of *religion* (understood *in modern terms*, as a separate realm of social and historical life). In Latin America "religion" and "populism" have long been indelible parts of an overarching representation of the region as backward, prone to authoritarianism and impervious to the impersonal, rational order of capitalism and modernity (Zanatta, 2008, 2014; Bastian, 2013; Oppenheimer, 2016; Giorgi, 2018). However, neither prevalent images stand the test of "really existing" Latin American contexts – whether national, local, transnational or diasporic. The latter at once blur neat distinctions stemming from "official" modern understandings of religion and the people and preserve the associative and mobilisation dimensions that both concepts evince (Montero, 2012; Burity, 2015a). Ambivalence shakes arguments based on intrinsic logics of capitalism and modernity as well as arguments based on metaphorical application of the two concepts that do not account properly for the performative, material force of their rhetorical ambivalence.

In the case of Pentecostalism, since its emergence in Latin America – a mixture of co-occurrence with international cases and the result of minority missionary initiatives – coming to be considered as part of the people involved a triple challenge. First, breaking the colonial association of national identity with Catholicism in all Latin American countries. Second, breaking the anti-popular prejudice of intellectual and political elites vis-à-vis movements of the poor and immigrants, which lacked international institutional support connections. Third, breaking the rejection of Pentecostal "sectarian" mores and rhetoric expressed through discourses favourable to syncretism and discourses of secularism. As a movement of ordinary people, focused on regionally peripheral populations, both ethnically diverse from the Ibero-Catholic matrix dating from colonial times (former black slaves and Central European economic migrants) and socially marginalised (rural and urban poor), Pentecostalism did not qualify as a religion "proper" for decades. Thus, a framework was constructed in which the postcolonial legacy does not actually establish an overcoming of coloniality, but its reinstatement, by means of an elitist and exclusionary national project also impacting on non-Catholic religions.

Until the late 1970s, Pentecostalism went unnoticed by many outside urban peripheries and the rural world. In this period, however, the various Pentecostal denominations and independent churches not only grew far above the historical Protestant denominations, they even penetrated the latter, via charismatic movements that initially split those denominations (mid-1960s to early 1970s) – forming "renewed" versions of them – and then resurfaced in them (as of the late 1980s) as internal currents. At the threshold of the 1980s' return to democratic rule in South America, Pentecostalism already accounted for two-thirds of all Protestants (with some national variations) and experienced an accelerated growth, regionally. The crossovers between the processes of democratisation and the evidence of numerical growth (among other factors, which I cannot develop here) catalysed a *minoritarian self-assertion* as of the mid-1980s. In the following decades this minoritisation went from claiming to be counted in the ranks of the national people to presenting themselves as proponents of a "reform" that ultimately amounts to a general Pentecostalisation of society, the constitution of a new people (Burity, 2016, 2017). I will explore, in the fourth section, the problematic aspects of this "reformist" proposal, not so much because they represent a feared confessionalisation of politics, but because they significantly transform the ethos and whatever might be called the Pentecostal view of politics and social life, towards a new articulation between moral and cultural conservatism, on the one hand, neoliberalism and xenophobic nationalism, on the other.

Ecumenical Protestants as harbingers of a new people

Pentecostals are not alone in their public emergence. Way before them, in the mid-1950s, the ecumenical movement had laid roots across Latin America, promoting a specific mobilisation of young clergy and laypersons around questions of piecemeal development or revolution (corresponding to increasingly jarring internal orientations). By the 1970s, ecumenicals had joined the main current of liberation theology, clashing with highly suspicious and refractory local churches and denominational structures and seeking solace in the shade of the Catholic church's pastoral networks. The practice of grassroots mobilisation, particularly in urban neighbourhoods and in rural areas, through social projects funded by international or global ecumenical sponsors, is connected to "popular education" methodologies inspired in the work of Paulo Freire, a Brazilian educationalist and former staff of the World Council of Churches, as an exile escaping the post-1964 military regime in Brazil (Bastian, 1993; Barreto Jr., 2012; Burity, 2015b, 2018b).

It is against this backdrop that a distinctive Latin American ecumenical discourse on the people developed, influenced by Catholic and Protestant liberation and political theologies and implemented within the milieu of Catholic pastorals or through local social projects, many times in total disconnect with Protestant churches, in the midst of poor communities. "The people" is virtually the same as the one constructed by Catholic social discourse: it the grassroots of society, those who have been abandoned by governments but are prepared to increasingly risk their lives to claim justice and peace. Through long-term involvement with international organisations belonging to the global ecumenical movement, ecumenical social activism sought to meander through the narrow spaces left for survival in churches not only basically bent on apoliticism-*cum*-subservience to authoritarian governments, but also fiercely anti-communist. In addition, such links would provide ecumenical leaders with authority and back-up to voice popular demands together with other religious and secular partners (Petrella, 2006; Bastian, 2013; Jones et al., 2014; Baptista, 2016).

The same temporality of democratisation provided fertile grounds for many ecumenical initiatives, particularly since the 1990s. This is when a combination of global ecumenical engagement with multilateral bodies, governmental international cooperation agencies in Europe and North America, and the emerging anti- and later alter-globalisation movement provided new momentum for local and global initiatives. The increasing strengthening of network ties between local, national and international organisations, transnational and global movements, in order to partner with national and local governments on public policies or specific programmes, created other platforms for ecumenical social activism.

That evolved as a new pattern of public engagement: partnerships, public policy and social movement networks, *glocal* mobilisations. A pluralistic understanding of the people brought ecumenicals nearer to emerging minority politics and farther away from Pentecostal preferred strategies of seeking influence through electoral politics and governmental coalitions. The discourse on inclusion, human rights and distributive policies embraced by ecumenical organisations and activists came into increasing tension with the traditional moral stance of most Pentecostals, who started a steady process of disassociation with incumbent left-wing and centre-left governments, leading to a final break and adoption of a fully conservative agenda and ethos by large swathes of Pentecostal churches and leaders (Huaco Palomino, 2013; Parker, 2016).

Minority peoples: an "evangelical-capitalist resonance machine" facing a socio-political-religious arena of pluralisation and struggles for justice?

After over a decade of so-called progressive governments across the region, there are strong signs of a concerted backlash engaging local and global elites and corporate actors against hard-won achievements in addressing renitent inequalities, which affirmed multidimensional rights (thus articulating citizens' rights to women, blacks, indigenous people, gays, and religious minorities' claims to recognition and justice). In this new scenario, the adversaries of capital are denounced as "populist", but the very discourse against those sounds a lot like a right-wing version of populism. Once again, conservative religion also falls under the denunciation of populism. Ecumenical radical articulation of the people as a plurality of social groups articulated through a common commitment to solidarity and justice also comes under fire from yet other critics of populism. Clearly, the problem identified by Laclau hits back: given such a cacophony of views and contents associated with populism, either one needs to conclude for the emptiness of the word or to employ a different approach.

Here we need to add a different track to Laclau's analysis, as presented above. Not all populist formations are formed on account of a situation of unequivocal oppression by an objectively given force. Populism – as a division of society in two antagonistic camps, giving rise to a new assertion of a people – can also emerge as a result of the effects produced by a popular and democratic hegemonic order on resentful beneficiaries of a *status quo ante*. This emergence may also be triggered by bad or ineffective handling of such reactions from privileged groups, and contingent crises which lead to failures to fulfil promises, meet demands or renew the existing order's bases. The question is how deeply those effects impact on privileged groups and under which conditions they may regroup to unleash an open antagonism to what appear as failures of the hegemonic order. Under such conditions, the heterogeneity of demands – from above (former privileged groups) and from below (ordinary people affected by economic crises) – can produce an equivalence among them as they come to face those effects as intolerable: *reactions* and *frustrations* (not simply the latter) can thus converge and set off a conservative or radical "return to populism" (Nadal et al., 2012).

This seems to me to be a suitable theoretical narrative to delineate, in broad strokes, the process of deconstruction of the coalition led by the

Brazilian Workers' Party, starting in 2013, with massive street demonstrations in which all sorts of grievances and demands were voiced. The double outcome of the process was the impeachment coup in 2016 and the election of an extreme right-wing candidate in the 2018 presidential election (Bringel et al., 2013; Velasco e Cruz et al., 2015).

In this process, we will find a new configuration of Pentecostal populism: its strong alignment with a coalition of economically ultraliberal, politically anti-popular, and culturally moralistic and racist forces. A dissolution of the Pentecostal minoritisation from within, through the neutralisation of its inclusive and demo-cratising energies (as a space for mobilisation, organisation and expression of millions of people on the basis of a religious formation) by a discourse of resentment and moral and political revenge, autoimmunitarian (that is, promoting an unqualified attack on groups deemed to be weakening traditional values and challenging authority) and undemocratic (Burity, 2018c; Vital da Cunha, 2018).

This new configuration implies the abandonment of the initial, 1980s' strategy of building a Pentecostal people as a plural, "multicultural" people, of which the Pentecostals would be part through co-belonging. Now, feeling both empowered by their participation at the core of the post-impeachment coalition and threatened by minorities among whom they once belonged, Pentecostal leaders moved towards a strategy of demonisation of pluralism and uncritical affirmation of a majority "Christian" identity, a "Judeo-Christian tradition" supposedly underlying Brazil's national identity. Ironically, such a tradition was historically fully conjugated according to the Roman Catholic grammar and only very reluctantly opened itself to "hosting" Pentecostalism.

The Pentecostal attempt to lead a right-wing populist coalition on two occasions – the beginning of Michel Temer's post-impeachment administration and Jair Bolsonaro's 2018 presidential campaign – has resulted in a painful referral of its ambitions back to their "due place": that of a religious minority with strong popular bases but without real power to represent the whole anti-Workers' Party reaction.

Despite attracting millions of people to its entrepreneurial, transactional and self-assertive spiritual rhetoric, post-1990 Brazilian Pentecostalism has not only disavowed any equivalence with the demands of the social actors above – organised around affirming citizenship rights (including equal treatment of all religions by the state) and tackling inequalities. It has also become once more the target of negative, discriminatory minoritisation suffered until the late 1990s. The closer Pentecostal leaders come to the far-right agenda – re-enacting a very similar version to Connolly's "evangelical-capitalist resonance machine" of the Bush era – the more they get exposed to the vigilance and opposition of former sympathisers, allies or fellow travellers of the post-1980s democratising project, now turned bitter enemies.

Although still "surfing" the ebbing conservative tide, the mandarins of the Pentecostal political and pastoral elite have faced the fragility of their power base and been confronted with the unacceptability of their call for social reordering, based on traditional moral criteria and untrammelled neoliberal deregulation and privatisation (Frente Parlamentar Evangélica, 2018; Fábio, 2018). They thus come to oppose the entire cohort of actors who had equally benefited from the expansive inclusion of the democratisation period: Afro-descendants, indigenous groups, women, LGBT groups, environmentalists, rural workers and various other social movements.

The bloc articulated in 2016 was partly redefined during the intensely polarised 2018 elections, and the first movements of the new government have deeply disappointed conservative Pentecostalism: despite their growth in Parliament, they have failed to achieve any corresponding success in terms of government positions. Their views are now openly challenged by the media and progressive social and political forces.

As for ecumenical social activism, the post-2013 juncture actually created new opportunities. It gained modest acknowledgement through the activation of ecclesiastical structures (historical churches, the National Council of Christian Churches, and the Catholic Bishops' Conference). It also mobilised the web of ecumenical NGOs and networks to make public statements, participate in demonstrations and protests and re-experience coordinating efforts to feed international organisations, news companies and alternative media with critical accounts of what has happened in Brazil under the new conservative wave. Having reinforced its links with all sorts of left-leaning movements, the ecumenical camp has heightened its more confrontational, "prophetic" ethos, serving as a mediating platform for the expression of grievances, complaints and claims from those defeated by the movement towards conservatism.

Final remarks

The postcolonial debate highlights the always-situated character of knowledge and calls for reflexive acknowledgement of one's positioning and belonging. Abandoning any pretence to universal or neutral knowledge, postcolonial authors – in Latin America they increasingly resort to a *decolonial* lexicon – make room for those forms of life, practices and wisdom that fell under the radar of colonial forms of description and analysis. This counts for subaltern actors, political forms and religion. What I have tried to do here is to hint at one possible articulation of these dimensions, by (a) accepting provisionally the association in large segments of academic and political discourse between ordinary people's subalternity and populism as a form of domination or a naïve expression of people's voice that makes little of the formal institutions of power and decision-making; (b) noting the frequent connection of subaltern, deficient forms of relating to/making sense of historical and contingent circumstances through enchanted, religious categories; (c) exploring, based on Laclau's theory of populism (itself a decolonising gesture, as it reverses the notion of a deficient form of politics to argue for a general theory of politics), the dense texture of this relationship between religion and populism.

Although I have focused strictly on the Latin American scenario, bringing the Brazilian case to the fore, it would not be difficult to extend many aspects of the narrative to situations and trends already at work in Europe and North America. And not just for the presence of Latin American migrants or diasporas. The increasingly intricate links between local experiences and faraway dynamics and flows produces simultaneity and difference and opens the possibility for a double operation: (a) the reversal whereby categories and practices originating in subaltern or peripheral societies and by subaltern actors come to be deployed to understand and explain processes in the supposedly more advanced societies; (b) the worrying developments leading to perceived forms of violence, growing inequalities and weakening of democratic institutions and culture are no longer a remote object of observation for North Atlantic societies; likewise, the emergence of alternatives to these troubling developments no longer affords spatial distances and hierarchies, but circulates and can well serve the needs of diverse social settings.

In our case, religious minorities – both Pentecostal and ecumenical – have engaged in processes of self-assertion (positive minoritisation) in order to resist and undo age-old forms of invisibility and repression (negative minoritisation). In engaging such processes those minorities have either had to reshape the very definition of the people in order to make room for their singularity or joined wider expressions of articulation of demands that have demarcated boundaries between those at the top and those at the bottom, the privileged ones and the excluded ones, fighting monopolies or syncretic forms of absorption and silencing subaltern religions, cultures, and identities. The story is open-ended, as are the tasks of creating new spaces for other forms of arguing, narrating and analysing.

References

Anderson, A. (2004) *An Introduction to Pentecostalism: Global Charismatic Christianity.* Cambridge: Cambridge University Press.

Arditi, B. (2014) 'El populismo como espectro de la democracia, respuesta a Canovan', in *La política en los bordes del liberalismo: Diferencia, populismo, revolución, emancipación*. Barcelona: Gedisa, pp. 107–158.

Baptista, P. A. N. (2016) 'Pensamento decolonial, teologias pós-coloniais e teologia da libertação', *Perspectivas Teológicas*, 48(3), pp. 491–517.

Barrera Rivera, D. P. and Pérez, R. (2013) 'Evangélicos y política electoral en el Perú. Del "fujimorato" al "fujimorismo" en las elecciones nacionales del 2011', *Estudos de Religião*, 27(1), pp. 237–256.

BarretoJr., R. (2012) 'The Church and Society Movement and the Roots of Public Theology in Brazilian Protestantism', *International Journal of Public Theology*, 6, pp. 70–98.

Barros, S. (2009) 'Salir del fondo del escenario social: Sobre la heterogeneidad y la especificidad del populismo', *Pensamento Plural*, 5, pp. 11–34.

Bastian, J.-P. (1993) 'The Metamorphosis of Latin American Protestant groups: A Sociohistorical Perspective', *Latin American Research Review*, 28(2), pp. 33–61.

Bastian, J.-P. (2013) *Protestantismos y modernidad latinoamericana: Historia de unas minorías religiosas activas en América Latina*. México: Fondo de Cultura Económica.

Bergunder, M. (2009) 'Movimiento pentecostal en América Latina: Teorías socio-lógicas y debates teológicos', *Cultura y Religión*, 3(1), pp. 6–36.

Biglieri, P. (2017) 'Populismo e emancipações: a política radical hoje, uma aproximação (com variações) ao pensamento de Ernesto Laclau', in D. Mendonça, L. P. Rodrigues, and B. Linhares (eds) *Ernesto Laclau e seu legado transdisciplinar*. São Paulo: Intermeios, pp. 19–38.

Bringel, B., Benzaquen, G., Alcântara, L., and Gomes, S. (2013) *As Jornadas de Junho em perspectiva global* (NETSAL Dossiers, 3). Rio de Janeiro: IESP. Available at: www.academia.edu/10068329/_2013_As_Jornadas_de_Junho_em_perspectiva_globa l [Accessed 16 January 2019].

Burity, J. (2015a) 'A cena da religião pública: contingência, dispersão e dinâmica relacional', *Novos estudos CEBRAP*, 102, pp. 93–109.

Burity, J. (2015b) 'Políticas de minoritização religiosa e glocalização: notas para um estudo de redes religiosas de ativismo socio-político transnacional', *Revista Latinoamericana de Estudios sobre Cuerpos, Emociones y Sociedad*, 18(7), pp. 9–30.

Burity, J. (2016) 'Minoritization and Pluralization: What Is the "People" That Pentecostal Politicization Is Building?', *Latin American Perspectives*, 43(3), pp. 116–132.

Burity, J. (2017) 'Authority and the In-common in Processes of Minoritisation: Brazilian Pentecostalism', *International Journal of Latin American Religions*, 1(2), pp. 200–221.

Burity, J. (2018a) 'Espiritualidades del consumo, del resentimiento y del agonismo político: religión pública versus desinstitucionalización religiosa', in J. Cruz Esquivel and V. Giménez Béliveau (eds) *Religiones en cuestión: campos, fronteras y perspectivas*. Buenos Aires: Ciccus, pp. 283–308.

Burity, J. (2018b) 'Formação, convencimento e mobilização: a construção do povo nas instituições e redes ecumênicas', in A. C. Lopes, A. L. A. R. M. de Oliveira and G. G. S. de Oliveira (eds) *A teoria do discurso na pesquisa em educação*. Recife: EdUFPE, pp. 361–402.

Burity, J. (2018c) 'A onda conservadora na política brasileira traz o fundamentalismo ao poder?', in R. de Almeida and R. Toniol (eds) *Conservadorismos, fascismos e fundamentalismos: Análises conjunturais*. Campinas: EdUnicamp, pp. 15–66.

Connolly, W. E. (2008) *Capitalism and Christianity, American Style*. Durham: Duke University Press.

Corten, A. and Marshall-Fratani, R. (eds) (2001) *Babel and Pentecost: Transnational Pentecostalism in Africa and Latin America*. London: Hurst & Co.

De Cleen, B., Glynos, J., and Mondon, A. (2018) 'Critical Research on Populism: Nine Rules of Engagement', *Organization*, 25(5), pp. 649–661.

Dempster, M., Klaus, B. D., and Petersen, D. (eds) (2011) *The Globalization of Pentecostalism: A Religion Made to Travel*. Eugene, OR: Wipf and Stock.

Fábio, A. C. (2018) 'As propostas da bancada evangélica, em 4 linhas centrais', *Nexo*, 11 November 2018.

Finchelstein, F. and Urbinati, N. (2018) 'On Populism and Democracy', *Populism*, 1, pp. 15–37.

Frente Parlamentar Evangélica (2018) 'Manifesto à Nação: O Brasil para os brasileiros', [online]. Available at: https://static.poder360.com.br/2018/10/Manifesto-a-Nacao-frente-evangelica-outubro2018.pdf [Accessed 10 March 2019].

Freston, P. (ed.) (2008) *Evangelical Christianity and Democracy in Latin America*. Oxford: Oxford University Press.

Giorgi, J. (2018) 'El populismo religioso se inserta en la política de América Latina. *El Observador*, [online], 1 February 2018. Available at: https://www.elobservador.com.uy/

nota/el-populismo-religioso-se-inserta-en-la-politica-de-america-latina–201821500 [Accessed 25 March 2019].

Huaco Palomino, M. A. (2013) *Procesos constituyentes y discursos contra-hegemónicos sobre laicidad, sexualidad y religión: Ecuador, Perú y Bolivia*. Buenos Aires: Clacso.

Jones, D., Luján, S. and Quintáns, A. (2014) 'De la resistencia a la militancia: las Iglesias evangélicas en la defensa de los derechos humanos (1976–1983) y el apoyo al matrimonio igualitario (2010) en Argentina', *Espiral - Estudios sobre Estado y Sociedad*, 20(59), pp. 109–142.

Laclau, E. (2005) *La razón populista*. México: Fondo de Cultura Económica.

Laclau, E. (2014) 'Why constructing a "people" is the main tasks of radical politics', in *The rhetorical foundations of society*. London: Verso, pp. 139–179.

López Rodriguez, D. (2014) *La seducción del poder Los evangélicos y la política en el Perú de los noventa*. Lima: Ediciones Puma.

Medina, N. and Alfaro, S. (eds) (2015) *Pentecostals and Charismatics in Latin America and Latino Communities*. New York: Palgrave Macmillan.

Montero, P. (2012) 'Controvérsias religiosas e esfera pública: repensando as religiões como discurso', *Religião & Sociedade*, 32(1), pp. 167–183.

Márquez Restrepo, M. L., Pastrana Buelvas, E. and Hoyos Vásquez, G. (eds) (2012) *El eterno retorno del populismo en América Latina y el Caribe*. Bogotá: Editorial Pontificia Universidad Javeriana.

Oppenheimer, A. (2016) 'Populismo religioso en América Latina'. *El Periódico*. [online], 11 November 2016. Available from: https://www.elperiodico.com/es/op inion/20061216/populismo-religioso-en-america-latina-5409594 [Accessed 22 March 2019].

Parker, C. (2016) 'Religious Pluralism and New Political Identities in Latin America', *Latin American Perspectives*, 43(3), pp. 15–30.

Pérez Guadalupe, J. L. (2017) *Entre Dios y el César: El impacto político de los evangélicos en el Perú y América Latina*. Lima: Konrad Adenauer Stiftung/Centro de Estudios Social Cristianos.

Petrella, I. (2006) *The Future of Liberation Theology: An Argument and Manifesto*. London: SCM.

Velasco e Cruz, S., Kaysel, A. and Codas, G. (eds) (2015) *Direita, volver! O retorno da direita e o ciclo político brasileiro*. São Paulo: Fundação Perseu Abramo.

Vital da Cunha, C. (2018) *Religiões, sentimentos públicos e as eleições 2018* [online]. Rio de Janeiro: ISER. Available at: https://br.boell.org/pt-br/2018/08/27/reli gioes-sentimentos-publicos-e-eleicoes-2018 [Accessed 28 January 2019].

Wynarczyk, H., Tadvald, M. and Meirelles, M. (eds) (2016) *Religião e política ao sul da América Latina*. Porto Alegre: CirKula.

Yong, A. (2010) *In the Days of Caesar: Pentecostalism and Political Theology*. Grand Rapids, MI: Wm. B. Eerdmans.

Zanatta, L. (2008) 'El populismo, entre religión y política: Sobre las raíces históricas del antiliberalismo en América Latina', *Estudios Interdisciplinarios de América Latina y el Caribe*, 19(2), pp. 29–44.

Zanatta, L. (2014) 'La relación entre populismo y religión salta a la vista. La Voz', [online], 10 July 2014. Available from: https://www.lavoz.com.ar/ciudad-equis/la-rela cion-entre-populismo-y-religion-salta-a-la-vista [Accessed 12 March 2019].

2 Parody, satire and the rise of populism under postcolonial criticism

A French and an Italian case

Cécile Vermot and Maica Gugolati

Introduction

"I no pay rent. [...] We no pay rent. Come on! We are niggers." "Integration is something that is important. My uncle, my father, twenty years in France and they never got integrated. Never. My father especially. Always dressed the same. Boubou,[1] jacket, sandals." The first statement comes from the song "Non Pago Affito" (2016) by the Italo-Ghanaian rapper Bello Figo feat. The GynoZz. The second one comes from "Mon père et l'intégration en France" (2012) by Patson, a French comedian from the Ivory Coast. Within their performances, these two artists present a voluntary re-enactment of racist and sexist stereotypes under the guise of satires and parodies of different types that troll contemporary European societies specifically on the issues of immigration and refugees. Stereotypes are recognisable characteristics that reduce a person or a category of population to those generalised traits, exaggerating and simplifying them (Hall, 1997). The mimicry performed by Bello Figo and Patson, as an adaptive behaviour (Taussig, 1993), is based on the enactment of generalised racist and colonial orientalist patterns (Bhabha, 1983; Said, 1979; Gugolati, 2018; Radio24, 2 January 2012), frequently addressed to immigrants by a section of the hosting countries where they live, Italy and France. More specifically, through their postcolonial performativity, they enact different stereotypes supported by the right-wing populist political, social and media discourse present in these two countries.

Despite the differences in political, social and economic context between Italy and France, there are similarities in the rise of right-wing populist parties and their rhetoric. In Italy 2018, the coalition between the anti-establishment Five Star Movement and right-wing Northern League party (BBC, 10 May 2018) brought them to power. In France, the election of Emmanuel Macron in 2017 took place after a confrontation in the second round between his party and the populist right-wing Rassemblement National.[2] In terms of the rhetoric, in both countries there has emerged a political lexical field based on "security", "nation" or "the people". Populist rhetoric can use emotional appeals (Mazzoleni, 2003, 2014; Demertzis, 2006; Bos, Van der Brug and de Vreese, 2011; Hameleers et al.,

2017; Rico et al., 2017; Wike et al., 2018). This rhetoric is commonly based on anger against an "elite" and the establishment, or a fear of "others", personified mainly by the figures of migrants.

This essay aims to analyse the tensions in the rise of populist rhetoric through satire and parodies performed by the two migrant artists. More specifically, it analyses to what extent racist and sexist stereotypes are performed through ethnic humour.[3] The comparison of postcolonial performativities of these artists allows us to deconstruct and show similarities and differences in right-wing populist rhetoric, In Italy and France the right-wing populist parties share a rhetoric that instrumentalises the issue of extra-European migration, mainly from the African continent and the Middle East, as "the problem" of the generalised condition for the European "crisis" (Inglehart and Norris, 2016).

To understand how stereotypes are transformed into satire and parody in order to denounce and question them, this essay will first highlight the link between right-wing populism and emotions. After explaining the methodology, it will define parody and satire through the cases of the artists Bello Figo and Patson. Third, it will highlight three main axes that carry and explain the entanglement between populist rhetoric and the postcolonial register, developing them in different terms.

The rhetoric and emotions of right-wing populism

The notion of populism is defined as an ideology that pushes a homogeneous social group against a set of elites who become dangerous "others" and are together depicted as depriving (or attempting to deprive) the sovereign people of their rights, values, prosperity, identity and voice (Albertazzi and McDonnell, 2008). To function as a political movement, populist ideology is accompanied by a rhetoric that mobilises ordinarily marginalised and excluded social sectors into publicly visible and contentious political action, while articulating an anti-elite, nationalist rhetoric that valorises the construction and the imagination of "ordinary people" (Jansen, 2011). The notion of "the people"[4] reflects an anti-status-quo discourse that refers to this category as the underdogs, the powerless, against the others: the powerful. This responds to the notion of the sovereign people as an identified collectivity, in antagonistic relations with an established order (Panizza, 2005).

The emotions of fear, anger and nostalgia are the ones that serve for the construction and affirmation of this specific rhetoric. They are not individual, but are felt and shared by groups in which people tend to self-categorise as members. Those who support right-wing populist movements usually think their country was better in the past than it is today. Nostalgia comes from the Greek verb *nostos*, which means "to return" and refers to the noun "pain". It is an emotion driven by a physical separation, whether geographical or temporal. To develop, it needs a past always idealised. This idealisation, which can be seen as a defence against loss, means that nostalgia is linked to conditions of both sadness and joy. In social science, it can be defined as a strategy to belong to the community continuing to participate in social life

(Anwar, 1979; Vermot, 2014). The sense of nostalgia may also have been reactivated by the prolonged European economic precariousness of today, which led to a generalised sense of loss of hope for a better future (Salmela and Scheve, 2017). Right-wing populist rhetoric represents an attempt to forge solidarity among "the people" through a rhetorical invocation that is based on an emotional communication, evoking a sense of nostalgia for a phantomic past that is always better than the present, as a way of facing the uncertainty of the future (ibid.). Nostalgia is linked with ethnocentrism, which appears as the second most important concern (Stokes, 2018). The emotions of fear and impotence evoked by the political, economic and social system are adopted by populist rhetoric, which finds scapegoat figures: the migrants, in order to exorcise this emotional charge.

This consequently creates a discourse of authenticity of suffering, that consists of a sense of imaginary victimhood (Giglioli, 2014) which wishes to gain authority based on a myth of suffering attached to the past. It functions as a subaltern claim for a dominant position that aims to legitimise its own position. Moreover, in the structure of populist rhetoric, the demagogic values celebrated by "the people" are announced and carried by a single leader figure. For the populist parties, "the people" support their leader who usually does not share their social class but belongs to a wealthier one to which "the people" aspire (Mudde, 2004). By so doing, there is the creation of a homogeneous demagogic in- and out-group discourse, where the sense of equality is translated into the notion of sameness (Canovan, 1999), allowing those who feel ignored and dismissed by the elites to gain a political voice and social visibility. These specific emotions shape the populist rhetoric; the chapter will continue by analysing the artistic practices and the way they operate within and about the populist discourse.

Methodology

For this research, the analysis of a musical production by Bello Figo and a humorous sketch by Patson have been carried out by viewing their productions on the YouTube platform. These two works were selected for their antithetical discourses facing the right-wing populist rhetoric that is present nowadays. The interest of these two case studies is not simply the act of inversion of racist commentaries by showing a positive image of their own communities (Muñoz, 1999) or the communities that they feel empathetic with (such as African refugees and immigrants from the African continent); rather it is the act of embodying and representing those stereotypes while claiming them, through satire for Bello Figo and parody for Patson. The social media platforms are in these cases used to narrate stories of minority subjects, highlighting the fictionality of the constructions of the colonial/racist stereotypical categories themselves. However, it does not just reproduce frameworks of understanding, memories, narratives, ideas, stereotypes and symbols on a national level; it brings a wider global perspective to it (Orgad, 2012).

Postcolonial performativities: the cases of Bello Figo and Patson

Ethnic humour as satire and parody

There is no single definition of humour. Nonetheless, humour can be delimited through four facets. First, it has a link with transgression against the social order about what is perceived as taboo. Second, it can be dissociated from laughter and smiling. Third, it differs according to culture, historical periods, generations, social groups and gender. Fourth, sharing humour can create in-group solidarity and therefore a separation from the out-group. Regarding the themes, humour considers social issues like social resistance, political conflicts and ethnic/racial issues reflecting the social hierarchy. It can both maintain social order and negotiate social boundaries (Billig, 2005; Kuipers, 2008; Le Breton, 2018).

Bello Figo and Patson take over racist and sexist stereotypes through satire in the case of the former and parody for the latter. Satire exposes a subject to ridicule, often through exaggeration; it is not necessarily designed to make people laugh (Low and Smith, 2007). It relies on a rhetorical strategy of persuasion (Duval and Martinez, 2001) distorting the target's representation through absurdity. Unlike satire, parody is usually comic, it amuses, it ridicules and mocks (Low and Smith, 2007). It leads the audience to be aware of the artificiality of the issue debated through mimicry (Kiremidjian, 1969).

Both artists are connected with a policy that concerns emotions, that regulates social sensibilities. Bello Figo and Patson are immigrants and live in countries that lack postcolonial debates. They both use grotesque excess (Mbembe, 1992), the practice of laughter and the performance of re-enactment as particular tools in order to invert and subvert social power relations (Bakhtin, 1970; Stam, 1989). They embody, display and re-enact, in their different artistic practices, some traits of the imperial imagination (Smith, 2012) as a form of self-determination that is founded in the performative act itself. Moreover, they "re-write", in a media and scenic sense, their personal stories blended with the generalised and fixed stereotypes (Bhabha, 1983) assigned by the receiving European societies. Through satire and parody that use humour, they unmask the absurd and irrational side of populist discourse. Nonetheless, the audience should be familiar with the ethnic humour practices used by the artists; if not, for an audience that is unaware of them, they might be interpreted as an offensive act or as a literal confirmation of the statements themselves. Those who do not get the joke become part of the problem, which the satire and parody address, becoming also a source of amusement for those who do (Low and Smith, 2007).

The musical case of Bello Figo

> I no pay rent.
> [...] We no pay rent.
> Come on! We are niggers.

All my friends came by boat.
Swag boat, as soon as we arrived in Italy we got:
house, cars, hotties.

[...] I no make labourer.[5]

[...] I do not get my hands dirty because I am already black: I am a refugee.
[...] We want WIFI, WIFI, WIFI!
[...] I sleep in 4 stars hotel because I am beautiful, rich, famous and black

This is part of "Non pago afito", a lyric by Bello Figo, a stage name that literally means "handsome cool or hot guy". Born in Ghana, Bello Figo came to Italy in 2004 at the age of 12 with the status of refugee. He started to compose and sing songs on the media platform YouTube, reaching (in 2018) more than 23 million views. In order to retain his freedom of speech, he refuses to sign up to any record label. In this song, by reproducing and embodying racial, social and sexual xenophobic stereotypes Bello Figo reveals some of the prejudices of Italian society. He talks about the prejudices and media commentaries on the subjects of immigration and refugees, which are daily debated in national politics and by the media. Instead of using a carnivalesque inversion (Bakhtin, 1970), in his song Bello Figo creates a mirroring reflection of the same stereotypes that objectify him, by declaring them. Just as Foucault (1966 [1970]) showed that the viewer of an artistic work becomes part of the subject shown in the painting while looking at it, in Bello Figo's song there is no separation between the xenophobic stereotypes stated by the artist and those declared by populist discourse and by many media commentaries on social media platforms. There is an interplay between those who receive and show the stereotypes, and therefore between the performer and the viewers. There is a blurring of boundaries between the figure who is subjugated by xenophobic stereotypes and the one who voluntarily enacts them; between the one who declares them and the one who listens to them.

Butler (1999, 2005) states that the drag performer's imitation of gender reveals the fictitious categorisation of gender itself and deconstructs it, demonstrating its own fictionality. Similarly, Bello Figo's satirical act of excessive embodiment and enactment of xenophobic debates, rubs the populist audience's noses in the fictionality of their own judgements.

By using a media platform, he trolls the listeners and viewers in order to provoke a strong reaction from the audience while simultaneously reflecting their own prejudices. There is a lack of moralism in his performance of his music. As he declares in his interviews (*Vice* 2018), he mixes his own biographical story with those of people he feels empathy with, such as the minority African migrants living on Italian soil, while at the same time developing the prejudices on immigration discussed by Italian media. In the act of embodying them, he becomes simultaneously the subject and the object of those stereotypes, dismantling them in a nonsensical logic. As he does so, the audience is faced with a staged absurdity that reflects the fictionality of the racial prejudices themselves. Bello Figo, when stating "I am beautiful, rich, famous and black", integrates the notion of blackness and foreignness into the celebrity paradigm (Gabler, 1998) and into consumerism, opening up the nationalist sense of economic and social protectionism to a wider intercultural issue. He adds to it a globalised paradigm of the red-carpet phenomenon, which shows the Italian audience that the contemporary aspiration to fame (Danto, 2009) embraces a whole supranational pop culture. Declaring that migrants desire and participate in the process of upward mobility simply by the fact of holding refugee status works to outrage what the populist discourse claims should be reserved primarily for Italian citizens.[6]

The comedy theatrical case of Patson

> [Patson's father] "My ungrateful son, you don't like the police. You are not grateful. Huh, you're not grateful! Your first bracelet? You were 12 years old. Who offered it to you? The police! Your first digital photo, it wasn't even fashionable. Who offered it to you? The police! [Patson's father's voice repeating his son's statement] 'I don't like the police!?' [Patson's father] You're ungrateful!"

Patson (Kouassi Patrice Mian) was born in 1974 in Ivory Coast. He started his career as a comedian in 2006 within the Jamel Comedy Club.[7] This club promotes humourists from French ethno-racial minorities questioning the French/Republican assimilation model. These artists, Patson included, are revisiting the stereotypical figures of migrants or the children of migrants. Within his performance, Patson seeks to question the issue of social diversity, complexifying the social image of "minorities" through the "story of self" of a character in a "subordinate" position. He uses jokes to counter the permanence of imagery inherited from the colonial era and to fight against stereotypes and their mechanisms of exclusion that imply a certain model of "Frenchness"[8] (Quemener, 2014, p. 86). Here, Patson takes up the image of the African migrant by describing directly his father and indirectly himself to a public, which is implicitly understood as being representative of Frenchness. Within this configuration, the public is laughing at stereotyped migrants but also at themselves. In other words, Patson used a representational practice of self-estrangement,[9] in which representation acts as an invitation to the audience to detach itself from the common-sense conceptions of their lives in the national understanding (Orgad, 2012).

Postcolonial performativity and populist rhetoric in Italy and France

"Us" versus "them"

The populist rhetoric of the right is based on an anti-elite, nationalist register that valorises ordinary people. It privileges a universalistic discourse that goes against the standpoint theories (Haraway, 1988; Harding, 2008), which maintain that all knowledge is socially situated within a context. This homogeneous universal discourse supported by right-wing populist rhetoric involves a universal analytical framework rather than responding to an ad hoc, and therefore diverse and partial, discourse based on geopolitical contexts. The category of "other" is employed in the dialectic structure "self–other" (also "us–them") where "other is a subject excluded from a dominant in-group which constructs an out-group" (Staszak, 2008, p. 2). There is therefore a distinction between "them", who belong to an outgroup, and "us", who belong to an in-group.

In populist rhetoric the figure of out-group is filled by the figure of migrant. Moreover, right-wing populist rhetoric is interrelated with a specific "national narrative" construction. A nation, as an "imagined community" (Anderson, 1983), allows the notion of a group condition to be defined with a subjective identification. Following Anderson's patterns (1983), this emotional bond, which allows the construction of the imagined group of the nation, includes in part the rejection, through anger, of the figure of an out-group which, in populist discourse, refers to migrants, strangers and outsiders. These are often stereotyped, victimised, demonised and invested with negative meanings through discourses and images (Kind and Wood, 2001 and Van Dick, 2000, in Orgad, 2012). It is specifically this distinction between "them" and "us" that appears within populist rhetoric in Italy and France, and also within Bello Figo's and Patson's performances.

In January 2016, Bello Figo was invited onto a televised broadcast (*Vice*, 2 December 2016), which also had as guests the Member of the European Parliament and columnist Alessandra Mussolini and the Federal Secretary of the Lega party Matteo Salvini. Their reactions were predictably opposed to Bello Figo's songs. In his songs, Bello Figo uses plural pronouns and plural forms of the verbs in order to highlight his positionality of a specific "us", identifying himself with the subjects of his song: refugees. The reactions of A. Mussolini were based on moral disdain; she told him to get "to work", suggesting that his status as a singer was fake. When she once again interrupted him saying "go back to your country", attempting to make him feel alienated from the place where he legally lives, Bello Figo answered correctly, "This is my country", insisting on the acceptance of the presence of migration and of the Italo-migrant generations as Italian citizens. The parliamentarian, in keeping with populist rhetoric, mixed moralistic commentaries with political positionings and frivolous personal opinions on the aesthetic value of his work and his persona, all delivered in the same conversation and with the same tone of voice. Following this TV appearance, some of Bello Figo's concerts were cancelled because of neo-fascist threats.

Patson, in his performance, emphasises a distinction between the pronouns "us", which refers to migrants, and "you", which refers to the audience but also to a "them", a category of people different from him, the "French people". This is a reference to the policy of assimilationism supported by the French government. In the sketch, Patson points out that his father and his uncle are not assimilated. Assimilation[10] is a process, which can take place through economic or sociocultural steps. In his performance Patson clearly evokes these different issues. After twenty years in France, his uncle and his father are still not wearing appropriate clothing for the climate and the society they live in, they speak broken French and have strong accents. Patson plays both his father and the son (himself) at the same time by telling the story to the public. He translates the father's words to the public, to the "French people" who are presumed not to understand the African way of speaking French. By not knowing how to pronounce the host language, his father shows that he will never integrate in the country. Patson, using an ironic caricature of cultural difference, allows himself to evoke the excesses of a perpetually hosting French culture (Quemener, 2014).

The linguistic aspect of the supposed assimilation component has been used voluntarily by both artists, Bello Figo and Patson, in their mimicry of broken Italian and broken French.

Politically and historically, Italy and France have treated the racial issue differently. In Italy "racial laws"[11] were formally established during the Second World War and were cancelled by the constitution at the end of that period. Despite this legal change, in the current media discourse some politicians (*Panorama*, 5 September 2018), when discussing the phenomenon of immigration, have (erroneously and illegally) raised the question of a white race (referring to a generalised notion of Italian citizens). In France, the question of race was historically neglected (D. Fassin and E. Fassin, 2006). The French narrative is built around the myth that French citizens live free, equal and in a fraternal country, under the protection of human rights.[12] Past recurring social tensions have allowed the debate to emerge progressively. The recognition of racial discrimination, "visible minorities", the issue of institutionalised discriminations has increasingly been faced by contemporary French society. Nonetheless, racist stereotypes continue to exist in the media, in jokes, daily discourse, representations and, obviously, in right-wing populist rhetoric.

Both artists are playing with migrant stereotypes and populist rhetoric within their performances, but in keeping with their national context they treat the issue of blackness in different terms.[13] In Bello Figo's satirical song, he embodies and declares a generalised notion of blackness, referring to it mainly through the figure of the (black, male African) refugee who comes across the Mediterranean Sea, exactly as the populist politicians portray them. Patson, on the contrary, parodies the tensions between his father, who comes from an African country, and the host society to which he belongs, maintaining a separation between the subject (himself) and the object that embodies the stereotypes he mimics. For Bello Figo there is a homogeneity between the status of refugees with the etiquette of blackness, and for Patson the African continent, simplified as "Africa", is spoken of as a homogeneous borderless place, and not as a continent that is composed of different countries. It is the colour of the skin of both artists that serves as sufficient for a distinction of "othering" within the Italian and French societies. Bello Figo lists in his rap a series of xenophobic clichés that are spoken and shared on social media where "we" (black people, refugees, including himself) do not want to work as labourers, because "we" do not want to get dirty, since "we" are already "black". This part of the lyric, which has been here paraphrased, responds to the local general expression, which conflates being black with being dirty inverting and reusing it in his favour. On the contrary, Patson's humour is based on a caricature of the African migrant and himself (the son). Repeating the erroneous representations of a part of French society, he considers the African continent as a homogeneous country, since his skin colour serves as a sufficient element of distinctness from the French population.

Slackness[14]

Within their performances, both artists evoke slackness in different ways. Slackness can refer at the same time to negligence and can be a postcolonial and decolonial tool (Cooper, 2005). It transgresses and outrages, through explicit sexuality, the attitudes and statements imposed by Eurocentric values.

Within his song, Bello Figo uses slackness verbally and explicitly, which allows him to investigate how the "high/low" culture divides the audience. Women, or literally "pussy",[15] or, as he stated afterwards, "white pussies", are repeatedly mentioned during the song, in order to emphasise the machismo and antagonism with Italian manhood, or the "white male" as he defines it. Sexual stereotypes embodied by the racist colonial discourse are overstated, emphasising an over-sexualisation of the black male body (Fanon, 1952; Wallace, 1978).

Patson focuses on the issue of abuse of the system by a young unemployed migrant and his slackness as described by his father while observing him:

> [Patson's voice] "My father had heard in the neighbourhood that I did not like the police. [Patson imitates the accent of his father] 'Ehin? You do not like the police?' [Patson's voice] My dad, he loves the police. [Patson's father] 'Come here! What's your problem with the police, huh? I registered you at school.[16] [...] They fired you! I put you to work, they fired you! I put you in the assedic,[17] they fired you! You tell everyone you're doing the three eight![18] You sleep in the morning, noon and night. That's your three eight!'"

Patson, unlike Bello Figo, considers the bionomy between a taken-for-granted sense of gratitude to the host country and the uncomfortable disparity between that and the social reality. He does not treat ironically the issue of inadequacy felt by a migrant in relation to the assimilation system imposed by the society itself; rather, he exposes, through the views of his father, who experienced the colonial relationship between an African country and the "metropole",[19] the moral duty of migrants to be "thankful" and appreciative to French society for having welcomed them, offering them supposedly a better life. His humour extends to the relationship between "thankfulness" and the role of the police, a common discrimination issue for the majority of young citizens of the suburbs who are mostly from migrant backgrounds (Hargreaves and Perotti, 1993, in Orgad 2012). Through the mimicry of his father, he presents the general view of a young person living in the Parisian suburbs, who is associated in the social imagination with a delinquent.

> "[...] You listen to weird music: 'Murder the police, murder the police [with body gesture]. Woop woop. Murder the police. Woop woop'. If you continue, it's me that will kill you, even before the police [laughs]."

He is also addressing a political discourse specifically regarding slavery, post-colonialism, racism and discriminatory violence from the police. Within this part of the sketch, Patson is referring to a song from the iconic French movie *La Haine* (1995), about life in the Paris suburbs. In one scene, DJ Cut Killer plays the remix of the song "Sound of da police" (1993), by the by KRS-One, a North American rapper from Bronx, New York, from the French rap group NTM.[20] Both KRS-One and NTM are politicised rappers, who speak about slavery, postcolonialism, violence, and police beatings and racism. NTM explicitly expresses the persistence of institutional racism (Béru, 2008). Even though the rap group is no longer active today, the words "woop, woop, assassins de la police", which were incorrectly translated[21] and wrongly attributed to the French group, are still used today as a rallying cry during French demonstrations (*France Culture*, 2018). Through his sketch, Patson ironically denounces two issues at the same time: the violent systemic condition of people of colour in the suburbs, and the colonial hierarchical relations between postcolonised citizens and European countries, as personified by his father.

The fear of invasion

For the populist parties of the right migrants, as an identified "object" and "subject", may embody responsibility for the nation-state's instability. This can explain the fact that what seems to affect contemporary politics the most is the perception of danger based on the issue of immigration, rather than the actual rate of immigration itself (*Il Sole 24 Ore*, 29 August 2018). Europe today is facing political, economic and social uncertainty. In this situation, people feel or perceive a danger without being able to identify it, making them anxious. Uncertainty pushes people within an iterative decision process into a state of permanent alert, which leads to anxiety (Delumeau, 1978; Callon et al., 2001; Salazar, 2010). Through the right populist rhetoric, this anxiety is channelled into fear of the figure of the migrant[22]. The emotion of fear is linked to a situation of risk, where people put themselves on an episodic alert and express fear towards a danger that is likely to occur (Beck, 2001). Its instrumentation, its invocation and its political uses are no longer confined to the regimes of terror but are also found within democratic regimes (Crépon, 2008).

In the song by Bello Figo, he lists a provocative trilogy that responds to the embodiment of the xenophobic and racist commentaries that are common in the Italian media and in popular discussions. He states that "his friends", referring to black refugees, when they dock from the illegal boats that cross the Mediterranean Sea, automatically receive a car, a house and women. Here he responds to the simple reversed phenomena of the carnivalesque (Bakhtin 1970), where the inversion of habits and discussions are fully and literally identified and embodied. By mentioning cars, he responds to the question of social status and to the general sense of social ascending desire of hierarchy. By mentioning houses, he raises an important event that happened during the specific timeline of the song: in 2016 a violent earthquake in the centre of Italy destroyed several small towns and caused more than 3000 forced evacuations (into tented camps and hotels) at the national expense (*La Repubblica*, 28 August 2018). At the same period, the government established a reception system for refugees who are hosted temporarily in unoccupied low-budget hotels. Bello Figo's statement responds directly to the xenophobic comments of feelings of injustice, from those who believed the Italian government should have prioritised Italians over the refugees, offering hotel accommodation to them instead. Comments from the populist right assert that the Italian families who were victims of the natural disaster were abandoned while, on the contrary, rooms in luxury hotels (*Libero Quotidiano*, 26 August 2016) with free Wi-Fi were given to refugees. The singer, exclaiming "WIFI" (provided by the luxury hotels) in the song, continues by stating that he wants to call his "friends from the boats" inviting them to come, in order to increase the (false) (*Sole 24 Ore*, 28 August 2018) perception of "invasion" by felt by Italians (*Il Fatto Quotidiano*, 27 August 2018). He embodies the xenophobic statements shared by the media by deliberately enacting the fictional xenophobic discourse. In the 2016 TV broadcast (*Dalla Vostra Parte*, 2 December 2016) the current (2018) Italian Minister of the Interior M. Salvini, who belongs to the separatist and xenophobic political party Lega, stated exactly the literal meaning of Bello Figo's parodic lyric. His failure to understand the practice of trolling and the parodying of society in the song led to Salvini's populist discourse being affirmed publicly and in the media. By taking Bello Figo's parody literally, the viewer, and in this case even the politician, sees and hears a confirmation of xenophobic judgements. For the politician, Bello Figo explains literally what the Italian nation really does: it hosts false refugees who are not escaping from war or conflict in their homelands and who are given a comfortable life thanks to the welcoming associations and national policies.

Patson touches on the issue of invasion, but in different terms. He talks about his overcrowded family, the polygamy of his father when he leaves his "four wives and 38 children at the train station so as not to pay for their tickets" (here paraphrased). He continues with the issue of being welcomed in Europe. Within his performance he affirms that new family members are hosted by his father, without mention openly if they have arrived legally or illegally:

"Once, I was in my room. Well my room is also the room of my eight sisters [laughs], my eight brothers [laughs] and my three cousins who came from home[23] [laughs]."

Moreover, he mobilises the colonial past with the issue of the overcrowded apartments lived by immigrants in France:

> "At home we had no bunk beds [laughs]. You're lucky. You have bunk beds. We were piled on top of each other [laughs]. I slept like that and my brother slept on my neck, directly [laughs]."

Here Patson alludes subtly to the slave trade, superposing it with the overcrowded places where contemporary immigrants tend to live. In this frame of his speech, "you" from "You're lucky" is addressed to the "French people" who typically live on a different level of comfort.

Conclusion

This chapter has aimed to deconstruct the entanglement between right-wing populist rhetoric and some practices of ethnic humour from Italy and France, two countries that lack major postcolonial debates. It has been demonstrated how populism is correlated with collective emotions and how they become political tools even when faced with artistic practices that attempt to reveal the fictional construction of social attitudes and political ideologies.

Although Bello Figo is not a political singer, he became a political figure because he performed a reality that was taken to be political, as a result of his register and satirical methodology. His songs do not declare openly what he personally wants to say but re-enact the main populist discussion in Italy.

In the case of Patson, the use of self-estrangement allows him to perform French African migrant stereotypes, which are central to the right-wing populist rhetoric in France. Like Bello Figo, Patson does not openly position himself as an artist with a political discourse. Nonetheless, his performances indirectly question French society about racism and African stereotypes, echoing the populist rhetoric of the right.

These two case studies, enable us to show how contemporary right-wing populist rhetoric does not give space to a partial discourse (Haraway, 1988) that gives a voice to the diverse minorities living in the same country; neither it is open to an act of reflexive deconstruction of the issues, nor to the methodologies through which they are expressed. It seems that these contemporary societies do not follow the rapidity of social change increased by the phenomenon of sudden migration that diversifies, while encoding and decoding (Hall, 1973), the status quo of the artistic and entertainment performances of the nations. A study of the reception by different audiences of Bello Figo and Patson's performances should be carried out in order to understand how parody and satire can be used by populism in order to reinforce its own ideology and to allow the empowerment of these "minorities". Nonetheless, at this stage this research also describes how the artists use the practices of satire and parody as postcolonial tools to question racist and sexist stereotypes.

Notes

1 Boubou or bubu is West African dress worn by men and women.
2 Rassemblement National previously called Front National was created in 1972 by Jean-Marie Le Pen and is now led by his daughter Marine Le Pen.
3 Ethnic humour is related to racial competition and conflict, and deals with the social and cultural patterns that have arisen from them (Burma, 1946).
4 In nineteenth-century Europe, what was assumed to be "the people" was a social category identifying the society of intellectual and socio-economic inferiority. This modern notion of the underclass has medieval roots that associated it with unpredictable and brutal riots (Guerra, 1992). Under democratic systems, "the people" started to be identified as citizens. The modern notion of "the people" maintained ambiguously the medieval meaning of dangerous irrational "plebs". "The people" in the populist discourse are those who consider themselves to be disenfranchised and excluded from public life (Panizza, 2005).
5 The lyric is in broken Italian used in order to dramatise the Italian language as stereotypically spoken by immigrants.
6 The reference is indirectly to the American political slogan "America first" used by Donald Trump in his 2016 election campaign, which refers to a foreign policy that emphasises nationalism and protectionism. In Italy, a similar neo-patriotic statement is made by the Minister of Interior M. Salvini: "Italians first".
7 This institution was created by Jamel Debbouze, one of the most successful comedians in France, who is a child of migrants living in a suburb of Paris.
8 "Frenchness" is a neologism that appeared in 1963, simultaneously in Paris and Montreal. It was used in 1966 by Leopold Léopold Sédar Senghor to describe a communion of French language around the world (Vachon, 1968). It was only in the 1980s, in France, that Frenchness started to be used to speak about national identity (Vidal, 2009) often linked with the idea of whiteness, and of Gallic and Christian origins.
9 Estrangement is a symbolic process of distancing the audience from its own already established narratives, culture, politics and history.
10 It defines the disappearance of an ethnic/racial distinction and social differences homologising to the hosting one (Alba and Nee, 1997, p. 863; Rumbaut, 1997; Zhou, 1997).
11 They were established by the fascist government in 1938, published by the PNF, Partito Nazionale Fascista, 26 October 1938.
12 The human rights declaration initially excluded women and individuals from the colonies. Even today, human rights edict is translated into French as "droits de l'homme": man's rights.
13 According to by INED and INSEE, the first ground of discrimination expressed by "visible minorities" in France is related to skin colour (Beauchemin, Hamel, and Simon, 2018).
   Despite the Italian Constitution's article no. 3 and other more recent conventions which impose legal sanctions against ethnic, racial discrimination that punish who publicly proclaims fascist principles and methods, there is no explicit reference to discrimination based on language and skin colour in the Italian Penal Code (*ECRI*, 2016).
14 The term "slackness" defines a controversial aspect of West Indian, Jamaican dancehall music that has sexually explicit lyrics, performances and dances that outrage the middle class (Mordecai and Mordecai, 2001).
15 In the lyric the word used is "figa", a slang term that refers both to female genitalia and an attractive woman.
16 The way he pronounces the term "school" sounds like "alcohol".

17 The Assedic, Association for Employment in Industry and Commerce, now called Pôle employ, is a French employment agency.
18 In France, working on "three eight" is a shift-work scheduling system to ensure continuous operation 24 hours per day.
19 "Metropole" refers to the continent as opposed to the overseas territories.
20 NTM: Suprême Nique Ta Mère (literally: to screw your mother).
21 French listeners changed the American refrain to "killers of the police".
22 Within the sociology of emotions, fear, like all emotions, is defined as a social construction (Vermot, 2017).
23 In the original French version, Patson uses the term "au pays" which refers to an immigrant expression that refers to the notion of home as the African continent.

References

Alba, R. and Nee, V. (1997) 'Rethinking Assimilation Theory for a New Era of Immigration', *The International Migration Review*, 31(4), pp. 826–874.
Albertazzi, D. and McDonnell, D. (2008) *Twenty-First Century Populism: The Spectre of Western European Democracy*. New York: Palgrave Macmillan.
Anderson, B. (1983) *L'imaginaire national: réflexions sur l'origine et l'essor du nationalisme*. Paris: Éditions La Découverte.
Anwar, M. (1979) *The myth of return: Pakistanis in Britain*. London: Heinemann.
Bakhtin, M. (1970) *L'oeuvre de François Rabelais et la culture populaire au Moyen Age et sous la Renaissance*. Paris: Gallimard Edition.
Beauchemin, C., Hamel, C. and Simon, P. (2018) *Trajectories and Origins: Survey on the Diversity of the French Population*. Paris: Ined Population Studies.
Beck, U. (2008) *La société du risque: Sur la voie d'une autre modernité*. Paris: Flammarion.
Béru, L. (2009) 'Le rap français, un produit musical postcolonial?', *Volume! La revue des musiques populaires*, 6(1–2), pp. 61–79.
Bhabha, H. (1983) 'The Other Question ... Homi K Bhabha Reconsiders the Stereotype and Colonial Discourse', in F. Barker, P. Hulme, M. Iversen and D. Loxley (eds) *The Politics of Theory*. Colchester: University of Essex, pp. 18–36.
Billig, M. (2005) *Laughter and Ridicule: Towards a Social Critique of Humour*. London: Sage.
Bos, L., van der Brug, W. and de Vreese, C. (2011) 'How the Media Shape Perceptions of Right-Wing Populist Leaders', *Political Communication*, 28(2), pp. 182–206.
Burma, J. (1946) 'Humor as a technique in race conflict', *American Sociological Review*, 11(6), pp. 710–715.
Butler, J. (1999) 'Bodily Inscriptions, Performative Subversions', in J. Price and M. Shildrick (eds) *Feminist Theory and the Body*. New York: Routledge, pp. 416–422
Butler, J. [1990] (2005) *Trouble dans le Genre*. Paris: La Découverte.
Callon, M., Lascoumes, P. and Barthe, Y. (2001) *Agir Dans Un Monde Incertain – Essai Sur La Démocratie Technique*. Paris: Seuil. La Couleur Des Idées.
Canovan, M. (1999) 'Trust the People! Populism and the Two Faces of Democracy', *Political Studies*, 47(1), pp. 2–16.
Cooper, C. (2005) *Noises in the Blood: Orality, Gender, and the "Vulgar" Body of Jamaican Popular Culture*. New York: Palgrave Macmillan.
Crépon, M. (2008) *La Culture de la peur: Tome 1 Démocratie, identité, sécurité*. Paris: Editions Galilée.

Danto, A.C. (2009) *Andy Warhol*. Milan: Piccola Biblioteca Enaudi.

Delumeau, J. (1978) *La Peur en Occident, XIVe–XVIIIe Siècles*. Paris: Fayard.

Demertzis, N. (2006) 'Emotions and Populism', in S. Clarke, P. Hoggett and S. Thompson (eds) *Emotion, Politics and Society*. London: Palgrave Macmillan, pp. 103–122

Duval, S. and Martinez, M. La Satire (2001) 'Littératures française et anglaise', *Bulletin de la société d'études anglo-américaines des XVIIe et XVIIIe siècles*, 53, 261–266

Fanon, F. (1952) *Peau Noire, Masques Blancs*. Paris: Éditions du Seuil.

Fassin, D. and Fassin, E. (2006) 'De la question sociale à la question raciale?', in D. Fassin and E. Fassin, *Représenter la société française*. Paris: La Découverte, pp. 1–44.

Foucault, Michel. (1966) [1970]. *Les Mots et les choses: Une archéologie des sciences humaines*. Paris: Gallimard.

Gabler, N. (1998) *Life the Movie, How Entertainment Conquered Reality*. New York: Alfred Knopf Inc.

Giglioli, D. (2014) *Critica della Vittima*. Roma: Figure Nottetempo.

Guerra, F.-X. (1992) In *Modernidad e Independencias*. Bilbao: Editorial Mapfre.

Gugolati, M. (2018) 'Creation of an exportable culture: A cosmopolitan West Indian case', *African and Black Diaspora: An International Journal*, 11(3): Special Issue on Stuart Hall.

Hall, S. (1973) *Encoding and Decoding in the Television Discourse*. Birmingham: Centre for Contemporary Cultural Studies.

Hall, S. (1997) *Representation: Cultural Representations and Signifying Practices*. First Edition. Culture, Media and Identities Series. London: SAGE.

Hameleers, M., Bos, L. and de Vreese, C.H. (2017) 'Shoot the Messenger? The Media's Role in Framing Populist Attributions of Blame', *Journalism*, March, pp. 1–20

Haraway, D. (1988). 'Situated Knowledges: The Science Question in Feminism and the Privilege of Partial Perspective', *Feminist Studies*, 14(3), Autumn, pp. 575–599.

Harding, S. (2008) *Sciences from below: Feminisms, postcolonialities, and modernities*. Durham, NC: Duke University Press.

Inglehart, R.F. and Norris, P. (2016) 'Trump, Brexit, and the Rise of Populism: Economic Have-Nots and Cultural Backlash'. HKS Working Paper No. RWP16–026. Available at SSRN: https://ssrn.com/abstract=2818659 or http://dx.doi.org/10.2139/ssrn.2818659

Jansen, R.S. (2011) 'Populist Mobilization: A New Theoretical Approach to Populism.' *Sociological Theory*, 2 June.

Kiremidjian, G. D. (1969) 'The Aesthetics of Parody', *The Journal of Aesthetics and Art Criticism*, 28(2), pp. 231–242.

Kuipers, G. (2008) 'The Sociology of Humor', in V. Raskin (ed.) *The Primer of Humor Research*. Berlin: Mouton de Gruyter.

Quemener, N. (2014) *Le Pouvoir de l'humour: Politiques des Représentations dans les Médias en France*. Paris: Armand Colin.

Le Breton, D. (2018) *Rire: Une Anthropologie du rieur*. Paris: Editions Métaillé.

Low, B. and Smith, D. (2007) 'Borat and the Problem of Parody', *Taboo: The Journal of Culture and Education*, 11(1), Article 6.

Mazzoleni, G. (2003) *The Media and Neo-Populism: A Contemporary Comparative Analysis*. Santa Barbara: Praeger.

Mazzoleni, G. (2014) 'Mediatization and Political Populism', in F. Esser and J. Strömbäck (eds) *Mediatization of Politics: Understanding the Transformation of Western Democracies*. London: Palgrave Macmillan, pp. 42–56.

Mbembe, A. (1992) The Banality of Power and the Aesthetics of Vulgarity in the Postcolony', *Public Culture* 4(2), pp. 1–30.

Mordecai, M. and Mordecai, P. (2001) *Culture and Customs of Jamaica*. Westport: Greenwood Press.

Mudde, C. (2004) 'The Populist Zeitgeist', *Government and Opposition* 39(4), pp. 541–563.

Muñoz, J.E. (1999) *Disidentifications. Queers of Color and the Performance of Politics*. Minneapolis: University of Minnesota Press.

Panizza, F. (2005) *Populism and the Mirror of Democracy*. New York: Verso.

Rico, G., Guinjoan, M. and Anduiza, E. (2017) 'The Emotional Underpinnings of Populism: How Anger and Fear Affect Populist Attitudes', *Swiss Political Science Review* 23(4), pp. 444–461.

Rumbaut, R.G. (1997) 'Paradoxes (and Orthodoxies) of Assimilation'. SSRN Scholarly Paper ID 1881265. Rochester, NY: Social Science Research Network. Online at: https://papers.ssrn.com/abstract=1881265. [Accessed 19 April 2019].

Salazar, M. (2010) *Arquitectura política del miedo*. Buenos Aires: Elaleph.com.

Salmela, M. and Von Scheve, C. (2017) 'Emotional Roots of Right-Wing Political Populism', *Social Science Information*, 56(4), pp. 1–42.

Said, E. (1979) *Orientalism*. New York: Vintage Books.

Smith, L.T. (2012) *Decolonizing Methodologies, Research and Indigenous People*. London: Zed Books.

Stam, R. (1989) *Subversive Pleasures. Bakhtin, Cultural Criticism and Film*. Baltimore: Johns Hopkins University Press.

Staszak, J.-F. (2008) 'Other/Otherness' in R. Kitchin and N. Thrift (eds) *International Encyclopaedia of Human Geography*. Amsterdam: Elsevier. Online at: https://www.unige.ch/sciences-societe/geo/files/3214/4464/7634/OtherOtherness.pdf [Accessed 7 September 2018].

Taussig, M. (1993) *Mimesis and Alterity: A Particular History of the Senses*. New York: Routledge.

Orgad, S. (2012) *Media Representation and the Global Imagination*. Global Media and Communication. New York: Polity.

Vachon, G-A. (1968) 'La "Francité"', *Études françaises*, 4(2), pp. 117–118.

Vermot, C. (2014) 'Los sentimientos de pertenencia a la nación de los inmigrantes argentinos en Miami y Barcelona', in *Migración 2.0 Redes sociales y fenómenos migratorios en el siglo xxi*, Guadalajara: Universidad de Guadalajara Centro Universitario de Ciencias Sociales y Humanidades, pp. 30–45.

Vermot, C. (2017) 'Peurs et aspiration à l'émigration à Miami et à Barcelone des Argentins de la classe moyenne (1999–2003)', *Amérique Latine Histoire et Mémoire: Les Cahiers ALHIM*. Online at: http://journals.openedition.org/alhim/5815. [Accessed 1 October 2018].

Vidal, C. (2009) 'Francité et situation coloniale', *Annales: Histoire, Sciences Sociales* 64(5), pp. 1019–1050.

Wallace, M. (1978) *Black Macho & the Myth of the Super-Woman*. New York: The Dial Press.

Wike, R., Simmons, K., Silver, L. and Cornibert, S. (2018) 'In Western Europe, Populist Parties Tap Anti-Establishment Frustration but Have Little Appeal across Ideological Divide', Pew Research Center Online at: https://www.pewglobal.org/2018/07/12/in-western-europe-populist-parties-tap-anti-establishment-frustration-but-have-little-appeal-across-ideological-divide/ [Accessed 19 April 2019].

Zhou, M. (1997) 'Segmented Assimilation: Issues, Controversies, and Recent Research on the New Second Generation', *International Migration Review*, 31(4), pp. 975–1008.

Online articles and videos

Alberto, M. (2018) 'Migranti, tutti i numeri dell 'invasione' che non c'è', *Il Sole 24 Ore*, 29 August. Online at: https://www.ilsole24ore.com/art/mondo/2018-08-28/migra nti-tutti-numeri-dell-invasione-che-non-c-e-122548.shtml?uuid=AE2SIHgF& refresh_ce=1. [Accessed 31 March 2019].

BBC (2018) 'Five Star and League: Italy populist leaders close to government deal', 10 May. Online at: https://www.bbc.com/news/world-europe-44066711. [Accessed 31 March 2019].

Dalla Vostra Parte (2016) 2 December. Online at: https://www.youtube.com/watch?v= A2AGymqiYZw&list=PLIxA–wu9-IO5n3MrubkBYYHa3S7hJshA&index=73&t= 0s [Accessed 31 March 2019].

Il Fatto Quotidiano (2018) 'Migranti, Italia è Paese con percezione più distorta in Ue. Presenze sovrastimate e ostilità maggiore di tutta Europa', 27 August. Online at: http s://www.ilfattoquotidiano.it/2018/08/27/migranti-italia-e-il-paese-ue-in-cui-la-percezio ne-e-piu-distorta-presenze-sovrastimate-e-ostilita-maggiore-di-tutta-europa/4583970/. [Accessed 31 March 2019].

Le Prince, C. (2018) 'Assassins de la police: Histoire d'une hallucination collective', France culture. Online at: https://www.franceculture.fr/musique/assassins-de-la-police-histoire-dun-slogan-ne-dune-hallucination-collective. [Accessed 19 April 2019]

Libero Quotidiano (2016) 'Terremotati in tendopoli, immigrati in hotel: perché gli italiani s'infuriano', 26 August. Online at: https://www.liberoquotidiano.it/news/italia/ 11953439/terremotati-tendopoli-immigrati-hotel-italiani-furiosi-amatrice-.html. [Accessed 31 March 2019].

Lorusso, E. (2018) 'Leggi razziali, 80 anni fa la nascita del razzismo di Stato in Italia', *Panorama*. 5 September. Online at: https://www.panorama.it/news/politica/leggi-ra zziali-italia-80-anni-razzismo-stato-shoah/. [Accessed 31 March 2019]

Luca, L. (2018) 'Terremoto, ricostruita meno di una casa su 10: E 3mila sfollati vivono ancora in albergo', *La Repubblica*. 24 August. Online at: https://www.repubblica.it/ cronaca/2018/08/24/news/terremoto_ricostruita_meno_di_una_casa_su_10_e_trem ila_sfollati_vivono_ancora_in_albergo-204803855/. [Accessed 31 March 2019].

Marino, V. (2016) 'Un commento dei momenti più illuminanti di Bello Figo a Dalla Vostra Parte', *Vice*, 2 December.https://www.vice.com/it/article/vd55pb/comm ento-momenti-memorabili-bello-figo-dalla-vostra-parte. [Accessed 31 March 2019].

Radio 24 (2012) 'Stereotipi Africani', 2 January. Online at: www.radio24.ilsole24ore. com/programma/luogo-lontano/stereotipi-africani-191422-gSLAVH00. [Accessed 31 March 2019].

Salvia, M. (2018) 'Abbiamo parlato con Bello Figo di Italia, politica e razzismo', *Vice*, 4 May. Online at: https://www.vice.com/it/article/ne9ajd/intervista-bello-figo. [Accessed 31 March 2019].

Stokes, B. (2018) 'Populist views in Europe: It's not just the economy'. Online at: https://www.pewresearch.org/fact-tank/2018/07/19/populist-views-in-europe-its-not-just-the-economy/ [Accessed 19 April 2019].

3 Neoliberalism and populism in Argentina

Kirchnerism and Macrism as the two sides of the same coin

Maximiliano E. Korstanje

Introduction

In August 2013, Argentine journalist Jorge Lanata coined the term *"la grieta"* (i.e. the crack) to denote a state of discrepancy and tension in politics as never before. Per Lanata, this crack not only separated friends and relatives, but also pitted Argentinian citizen against citizen. Not surprisingly, this type of divide and rule logic that traversed through the presidency of Cristina Fernandez de Kirchner opened the door to a fierce debate regarding the role of populism and its effects on journalism. In 2013, during the broadcasting of Martin Fierro awards, Lanata caustically alerted, far from being closed this *"grieta"* (crack) threatens the wellbeing and cohesion of Argentine society. From that moment onwards, the term *la grieta* became widespread through the media, on screens and in countless papers and books. To some extent, Kirchnerism and the Kirchnerites exploited an old squabble between academicians who proclaimed themselves the true knowers and the journalists, an underclass vilified by scholarship. This chapter centres on "populism" as the main theme, while other secondary themes are treated. I polemically hold that far from representing a real tension, *la grieta* seems to be an ideological narrative oriented to reorganising social ties after the stock and market crisis that whipped the country in 2001. The question is whether neoliberalism, when it arrived in the 1990s, came to stay. Neither Nestor and Cristina Kirchner nor Mauricio Macri[1] now abandoned the cage imposed by neoliberalism. Kirchnerism successfully gave to bit-players an important reason to be committed to politics, but the same neo-populist form of government was adopted by Macri just after he defeated Daniel Scioli in elections. Since 2001 the conditions of lay-citizens were not improved, reproducing not only serious material asymmetries but cultivating resentment and collective frustrations. Populism is, hence, sensitive to the climate of deprivations and uncertainty. Populism's pursuit of abstract goals moves individuals not only to believe that they are part of a historical process, but also a legacy for the next generations. This narcissistic discourse (similar to other reactionary romanticisms) paved the way for the rise of a new politics which looked to discipline journalists, as well as co-opting renowned academicians in what Jean Baudrillard dubbed the "great simulacra".

It is not surprising that the effects of romanticism in politics was widely studied and discussed by Hannah Arendt in her examinations on Nazism. She was an authoritative voice in the review of the complacency of many Nazi officials regarding the systematic assassination of vulnerable citizens. She coins the term "the banality of evil" to denote the indifferent behaviour of Adolf Eichmann once tried in Jerusalem. Of course, neither Kirchnerism nor Macrism can be compared to Nazi Germany, though both evolved according to the imposition of a cult that combines religiosity and secularism. One of the aspects that defines modern politics is the implantation of an existential doubt, which marks that the Other always lies, while I have the word of God in my hands. In the same way, Kirchnerism casts a radical doubt to undermine Clarin's credibility, Mauricio Macri creates a shadow of suspicions revolving around the Kirchnerites' corruption. Under the label of *"se robaron todo"* (in the office they ransacked everything), Macri's administration drew an ideological narrative that captured the already-existent frustrations of society. Both discourses emulate the logic of a neo-populism, which is encapsulated into neoliberalism.

The mythical struggle of Kirchnerism

On several occasions and in speeches, former president Nestor Kirchner affirmed his belief there "was no independent journalism", since journalists pursue corporate goals, subordinating themselves to the power of capital. In this way, business corporations pay for some journals for doctrinal and ideological support. This archetype was systematically replicated by the TV Program 6/7/8, a space hosted by the Official Television Channel. In this instance, originally panellists and reporters worked hard to unpack the ideological nature of journalism, which to date was considered a source of valid information. 6/7/8 discusses critically to what extent journalists do not transmit an ideological message to their audiences. Behind the journalists' message there lie some ideological explanations respecting events, which are not always adjusted to reality. In fact, the critical position of 6/7/8 posited an interesting discussion in the fields of journalism. However, over a number of years and once the presidency of Nestor Kirchner achieved peak popularity, this programme mutated towards more populist forms. It intended to make an uncritical defence of Nestor and Cristina Kirchner's presidencies while laying the foundations for "militant journalism" (*periodismo militante*), a type of partisan group (formed arbitrarily by journalists) who do not hesitate in expressing their political affiliations. 6/7/8, in this way, crystallised a sort of political witch-hunt against the opposition, as well as some critical colleagues. Even if Nestor Kirchner was originally on good terms with the Clarin Group (from 2004 to 2007), it is no less true that the dispute emerged when his successor, Cristina Fernandez de Kirchner, became annoyed after a cartoon ridiculed her following a polemical speech. In 2007, Romina Piccollotti, who was a subordinate of the former prime minister (Alberto Fernandez), filed a lawsuit against the Clarin Group around the so-called compliance during the

NRP (National Reorganisation Process) where the Juntas forced, killed and tortured thousands of Argentinian dissidents. To put this bluntly, *los desaparecidos* (the disappeared dissidents) were the "martyrs" of a process of extreme violence where terrorism wreaked havoc in the social imaginary. Clarin took advantage of its privilege position (with the Juntas) to gain further advantages in the media market. The Juntas allowed Clarin not only to enhance its considerable profits, but consolidated Clarin as the most powerful media corporation to date. The coming year witnessed an escalation of accusations that pitted Kirchnerism against Clarin.

Against this backdrop, it is important not to lose the sight of the following relevant facts:

- Some serious accusations were made about Ms Ernestina Herrera de Noble, former owner and director of Clarin, denouncing her son and daughter (Marcela and Felipe) for illegally expropriating during the dictatorship.
- La ley de medios (a media bill) was sanctioned by the Parliament aimed at disarticulating the power of Clarin in the market, forcing them to sell a part of their broadcasting enterprises.
- In the TV programme *A dos voces* hosted by TN (a Clarin subsidiary), the federal prosecutor Alberto Nisman denounced the current president Cristina Fernandez de Kirchner, the chancellor Hector Timerman and other top-ranking officials for concealing a pact of impunity with Iran, a country which firmly supported terrorists who perpetrated the attacks in Buenos Aires against the Jewish community in 1992 and 1994. Days after the denunciation was to be presented in the Parliament, Nisman was mysteriously assassinated in his home located in Puerto Madero, Buenos Aires. Needless to say, this event shocked public opinion.

Jean Baudrillard made a seminal contribution to the study of postmodernism and populism. From his viewpoint, the human senses are not based on direct experiences, but are mediated by "the gaze". The excess of visual simulation leads invariably to a culture of simulacra. To understand better this term, it is important to realise that events are enrooted in an historical past, in tradition, while simulacra produces "pseudo-events", which means events that never take place in reality. Since the sense of reality seems to be corrupted or at the best subordinated to the power of signs, as Baudrillard infers, the copycat replaces the original paving the way for the rise of an "order of sorcery". The postmodern world is not that different from the plot of Steven Spielberg's film *Minority Report*, where society successfully achieves a zero crime rate, but at a great cost: human free will. "Precogs" anticipate crimes before they are actually committed, enabling the government to initiate a programme of securitisation that curbs and eliminates crime from the streets. However, in the plot, the future overrides the present while the individual rights of citizens are vulnered in the name of security. In consonance with other voices such as Augé or Virilio, Baudrillard acknowledges that the hyper-reality reproduces "pseudo-events", which rule the present but never materialise in reality. Oriented to the future, politics, far from being the art of achieving, becomes "an aesthetical show" (Baudrillard, 2006) oriented to entertaining the lay-citizens. The condition of postmodernity works through mythical discourses and landscapes that are externally fabricated. As Paul Virilio puts it, the dictatorship of technology imposed a virtual reality, where the media says how events should be interpreted. Politicians maintain appearances, but at the bottom nobody knows their real intentions. Whereas the interactions in other epochs occurred in public space, now in post-modernity human existence has been emptied, leaving a gap which is filled by the media. (Virilio, 1995, 2005, 2010). John Armitage, one of the most pro-minent readers of Virilio, says:

> Perhaps the key to understanding the importance of Virilio's work on architecture, art and technology lies in the connection he makes between architecture, the organization of territory, and an idiosyncratic arche-ology of military fortifications, such as military bunkers, and the structure he creates for buildimng these connections with art and technology. Vir-ilio explains this structure as an archeology of military configurations, and it works less as chronology of western military fortifications and more as an aesthetic foundation for interdisciplinary cultural research.
>
> (Armitage, 2011, p. 7)

Similar remarks have been offered by Christopher Lasch, who in his seminal book *The Culture of Narcissism* accepts that the future of politics is grim in the same way professional politicians are not interested in making of this world a better placer, but only in pretending that everything is fine. As Lasch brilliantly observed, those cultures aimed at forgetting the past are doomed to repeat it, just here in the pre-sent. Starting from the premise that narcissism is a motion which alludes to an emotional dependence on the Other's view, Lasch argues convincingly that for the narcissistic character the world is nothing other than an extrapolation of its own desires. Not only are people moved by stories that stress the cult of hero, on its exceptionality as well as its mythical adventures in a struggle against the evil, but also politics, now distanced from the people, imposes arbitrarily a populist dis-course that make citizens believe "they are part of something important, a revolu-tion, a turning point in human or in the national history" (Lasch 1991: 37).

Intellectuals, democracy and the media

Whenever democracy is at stake, intellectuals and media play an active role in denouncing the arbitrariness of the executive branch. As Arendt eloquently explained, it is not an accident that those factors that facilitate often the rise of dictatorships are associated with a "banal evil" where citizens renounce their critical thought. Totalitarianism exhibits a radical evil, which consists in subordinating ethics to rational instrumentality, but "the banality of evil" signals a different process, where free citizens sacrifice their liberty to the bureaucracy of the state. In the midst of this mayhem, intellectuals may actively support undemocratic regimes when the opportunity arises. Beyond their erudition, many scholars and prominent writers have endorsed the figure of Adolf Hitler, motivated not by ignorance, but rather by the previous psychological frustrations that mutilated their social ties. Populism, in Arendt's view, does not correspond with irrationality but with a strong emotional poison aimed at revitalising the psychological deprivation of an alienated life (Arendt, 1964, 1972, 2013). Here two assumptions should be noted. On the one hand, Arendt was widely criticised for her notion of "the banality of evil". Based on a trial on Eichmann held in Jerusalem, Arendt portrayed a polemical image of middle-ranking officers of the SS, which ignited a fierce debate (Levinas, 2001; Peperzak, 2013). In other discussions, we have placed Arendt's argument under the lens of critical scrutiny. On the other hand, the subject may very well renounce its critical thinking but, in so doing, it is ethically responsible for such a renunciation. After all, it is not otiose to mention that Eichmann was an SS colonel; even he received specific training and of course was involved in anti-Semitic sentiment. Most probably, he was not a monster, but he should have been legally liable for his crimes (Korstanje, 2014). As Harry Frankfurt demonstrated, the subject is never dissociated from its ethical responsibility with the alterity (Frankfurt, 1987).

In line with the argument presented thus far, philosopher Mara Pia Lara (2007) calls attention to the impossibility of the subject in escaping its morality. Any genocide, as in Nazi Germany, opens the doors towards a moral tragedy. The derived trauma not only brings a psychological shock, but also allows a much deeper solidarity with the victims. Of course, as she contends, in the next years, the message can be politically manipulated or decontextualised according to some interests, but this does not mean that "remembrance" would not be vital in the process of resiliency.

Some voices alert us to the risk of demonising Nazis and other criminals as monsters or demons, simply because they are humans and should be tried in the same conditions as any other criminal. Demonising Nazis, a radical rupture enlarges human cruelty with the essence of evilness. To put this in bluntly, demons (as spiritual entities) cannot be judged by human law, while Nazis should be subject to a just trial (Morgan, 2001; Lang, 1990, 1996). This moot point leads Slavoj Žižek to claim adamantly that the Holocaust was commoditised to create a sentiment of victimhood, which today supports the ruling elite. Although Auschwitz was a terrible event, there were many others that were overlooked by Western nations (e.g. the genocide in Rwanda). This unfortunately suggests that the lives of some citizens are more significant than others. In Žižek's argument, academicians should not adopt a moral position revolving around crime. The monopoly of human rights is used to enhance profits (like the case of McDonald's and other business corporations which usufruct the other's suffering). Humanitarian assistance ushers society into a "false urgency" where the real reasons for disaster are replicated, never corrected (Žižek, 2002, 2004,

2005). In his recent book, *The Universal Exception*, Žižek (2015) makes a radical criticism against "liberal democracy and the nation-state". Democracies face serious challenges in struggling against populism and the arbitrariness of demagoguery. Žižek starts from the premise that the bipolar world, that of the US against the Soviet Union, has gone forever. Now, there is a clear dissociation between the subject of the enunciated and the subject of enunciation. This means that postmodern politics seems not to be different from the extortion of terrorists, where instrumentality prevails. This says, "I want you not only to do what I want, but I want to do it as if you really want to do it!" In this society, as Žižek adheres, citizens are imbued into an "extreme civility" that is oriented to expanding the individual's freedom, while at the bottom they are slaves of the "Big Other". As he writes:

> The role of civility in modern societies to the rise of autonomous free individual – not only in the sense that civility is the practice of treating others as equal, free and autonomous subjects, but, in a much more refined way, the fragile web of civility is the social substance of free independent individuals.
>
> (Žižek, 2015, preface, xv)

This extreme civility makes believe people they move freely when in reality they do not. Hence, Žižek argues that we live under the hegemony of totalitarian regimes camouflaged as "liberal democracies". In fact, he begs some more pertinent questions: what would happen if a political party, which is set to win an election, refuses the founding values of democracy? Is this emerging party punished or pressed to lead a clandestine life?

After all, liberal democracy rests on a formal legality, which is re-organised to create an adversarial game:

> Democracy ... concerns above all, formal legality: its minimal definition is the unconditional adherence to a certain set of formal rules which guarantee that antagonisms are fully absorbed not the agonistic game. Democracy means that whatever electoral manipulation takes place, every political agent will unconditionally respect the results.
>
> (Žižek, 2015, p. 59)

It is important not to lose sight of the fact that democracy and populism operate in a climate of tension between what Žižek calls "the obscene superego" and the "written law". He introduces the example of the Ku Klux Klan (the well-known American racist and white supremacist movement) whose actions are typified as illegal. While the formal law bans the KKK, prohibiting lynching against blacks, the KKK is implicitly legitimated by political power. Any new member has the prestige of belonging to this movement and this means he or she is hand-tied to report to the police if an Afro-American has been lynched. As Žižek alerts, prohibiting the behaviour without changing the cultural background leads to a cynical position. Is this part of what some scholars name the "the idolatry of human rights"?

Following Žižek's account, modern politics suggests the impossibility of human rights, which is subordinated to the hegemony of victimisation. To wit, Michael Ignatieff (2003) divides human rights as a question of politics from human rights as a matter of ideology. Ignatieff reminds us that the doctrine of human rights rests on shaky foundations because the violators are the same states that were originally created to protect the citizens' rights. In a critical essay-review, Korstanje (2016) discussed how populism tends to re-memorise an older trauma but evades the original background facts about how it happened. Kirchnerism and Kirchnerites – like many other populisms – recall a much-repressed trauma (like the crimes perpetrated by Juntas), but empty the discourse so as to legitimate their own ends. In the same way that Nazis remade the Treaty of Versailles as an "act of treason", Kirchnerism only tells a part of the real story. This begs some more than interesting questions: Is the media conducive to populism? Is media coverage the real reasons behind moral disasters, or offering a spectacle or a simulacrum?

In her work entitled *The Anthropology of News & Journalism*, Elizabeth Bird (2010) discusses to what extent there can be an "anthropology of journalism" without "a journalism of anthropology". The fact is that intellectuals deride journalists as naïve and superfluous, while journalists are proud of their capacity to sum-up complex stories in minutes or hours whilst avoiding academic jargon. Undoubtedly, both share the same goals of gathering and interpreting information to be deciphered by lay-people, while both agree to follow different methods.

Frank Esser and Jesper Stromback (2014) interrogated the role of media in established democracies not as the solution for people's problems, but as a form of legitimacy for partisan politics. Centred on the belief that media and politics should be dissociated, the specialised literature emphasises free speech as the touchstone of liberal democracy, but in fact the institutional rules, politics and media are rarely separated. Politicians manipulate journalism and intellectuals to gain further votes, in the same way that pressure groups bolster a fluid dialogue with politicians to gain further benefits:

> The basic idea behind the concepts of media logic and political logic is that media and politics constitute two different institutional systems that serve different purposes and that each has its own set of actors, rules and procedures, as well as needs and interests. These institutional rules and procedures can be formal as well as informal, and are often understood as the quasi natural way to get things done.
>
> (Essner and Stromback, 2014, p. 14)

Doubtless, there is a mediatisation of politics that creates great monopolies on both sides; journalists are often compromised when they should reveal vital information about the company for which they work. The autonomy of journalism, as well as the intelligentsia, with respect to the owners of capital, is a topic of great debate these days. Liberal democracy has serious problems in explaining how some news is distorted when it jeopardises the status quo (Essner and Stromback, 2014). Beyond these accusations, the importance of journalism depends on the accessibility of information given to lay-citizens. The paradox lies in the fact media and journalism buttress or undermine free speech and democracy with the same intensity (Blumler, 2014). In this context, as Mazzoleni (2014) holds, populist leaders capitalise on the discrepancies between journalists and politicians in their own favour. Populism needs media at a first stage, but once time elapses it puts media to the sword. To some extent, populism and media are inextricably intertwined. The need for instant attention leads populist politicians to not tackle the real problems in the economy, in which case, sooner or later, populism is confronted with countless unmet demands on the part of society. While paradoxically the popular leaders speak in the name of the population, there is a growing discontent generated by the excess of mediatisation (Mazzoleni, 2014). Still further, Claes de Vreese says that the efficiency of democracy in struggling against populism hinges on the autonomy the executive branch gives to journalism to offer a constructive feedback on the existent power: "While journalists and news organization may follow or deviate from a political elite actor's agenda, it is obvious that there is considerable leeway and autonomy on the side of journalism when deciding how to frame issues" (de Vreese, 2014, p. 148). But is democracy part of the problem or the solution?

In an incisive book, Matt Grossman (2014) questions the roots of American democracy. He toys with the belief that the fear of populism laid the foundations for a closed system, which is indifferent to citizens' demands. The efficiency of democracy rests on the division of powers, as well as numerous check-and-balance institutions that sometimes prevent social change. As he puts it, it is not true that the American system is immune to populism as some experts maintain; rather, power is centralised in the hands of a few autocrats, leaving a gap between citizens and their political institutions. Other scholars like Jonathan Simon (2007) hold that strong institutionalism, far from being part of the problem, impedes the rise of new populism that may harm the democracy. Over the years, different executive branches have attempted to create a discourse of fear in order to vulnerate the checks-and-balance powers. At the same time that some pressure groups may seek to create radical shift, the government invents an external enemy to deflect attention (Simon, 2007). In the next section, I shall review the classic textbook *On Populism* by Ernesto Laclau.

Ernesto Laclau (*On Populism*)

Ernesto Laclau needs little introduction; he has extensively theorised on the formation of reality and the role of media as distorters of reality. As one of the fathers of the Essex School, he focused on the effects of journalism in democracy and politics. Taking his cue from Freud, Tarde and Schmitt, Laclau explores the formation of semiotic paradigms, which subordinate the free will of citizens.

Populism is not a pathology of politics, it is politics itself. The core of modern politics is the efficacy in handling demands and processing them into forms of specific policies. The functioning of the system depends on how these demands are articulated or rejected. In order to keep their privileges, the ruling elite will handle as much as they can, but a portion of the demands will be unmet. This feeds-back a popular discontent, which will threaten the political elite at a later day. Laclau proffers a complex model that combines the chain of equivalence, the contagion effect in Gabriel Tarde, the theory of representations in Moscovici with the floating (or emptied) signifier proper of structuralism. Populism is not a social disease, as some American philosophers believe, nor the germination of a Spanish legacy in Latin America; instead, populism is understood as an instrument that forms the political ethos so as to revitalise the frustrations generated by the exploitation of capitalism. The workforce, if not alienated, is systematically exploited by capital-owners, who exert a radical hegemony (through journalism and the media) that draws the social imaginary. Paradoxically, while populism returns to hapless workers a certain psychological well-being, the term is vilified by privileged bourgeois groups as a symptom of social decline. For the specialised literature, populist leaders behave irrationally, embracing demagoguery as a form of politics. Laclau alerts us that this is a stereotyped vision of populism which has nothing to do with history. Part of the myopia, Laclau adds, of the experts in understanding populism comes from the derogatory nature they assign to the term. Elites often monopolise a closed system of signs, adjoined to symbolic resources to construct a net of understandings (of reality). Once some leaders make the decision to initiate a process of upward social mobility, the group is stigmatised as populist. Lebon and Tarde's prose shed the light on the functionality of imitation (or suggestion), which represents the oxygen of populism. Since interpretations are associated with signifiers, and of course the terms (enrooted in the language) are not objectively digested, Laclau writes that the reality is exhibited by the internalisation of external events, but always re-memorised according to the inner life. The signifier also can be emptied and filled in view of the subject's needs. In consonance with Schmitt, Laclau accepts that politics rarely are based on ethics, but on the levels of accumulated power to say what should or should not be done. This type of neo-decisionism divorces from ethics in the strict sense of the word. In consequence, populism reminds us that any social order centres on the influence of identification, which precedes identity. THEM are pitted against US *in a political dialectic*, where equivalence is equated to difference. This means that the discourse of equality leads to a rupture and fragmentation that causes material asymmetries. At the time populist leaders enter the scene, a chasm between low-ranked citizens and the oligarchy enlarges. Populism opens the door to a process of demarcation where the marks imposed by the elite are reverted.

Though Laclau was widely criticised for his polemical viewpoint, because of time and space, we shall avoid this discussion. Instead, we delve into the work of David Kelman (*Counterfeit Politics*) who has made a seminal contribution to the limitations of populism in modern politics. Kelman starts from the legacy of

Ricardo Piglia, an authoritative voice in the analysis of modern literature and the theory of conspiracy. For decades, the fathers of modern political science understood "the theory of conspiracy" as a pathological expression of politics, instead of what it is, namely the roots of power. Unless otherwise resolved, counterfeits politics take from secrecy the core to produce "a gap" which never can be empirically validated. This emptied core which is filled by the expectative of political leaders not only leads towards struggle with another contrasting group, but also enhances in-group cohesion. The credibility of leaders is founded in the impossibility of validating what they utter, while the silence produces a rupture between the official and unofficial story. To put this in other terms, at the time, the theory of conspiracy (plot) is tilted at producing a double-edged story where two sides are pitted against each other. Thus conspiracy seems not to be a symptom of corruption, but the necessary platform with which one discourse sets the pace for others. In his terms, politics offer an illusory state of emergency, where the sense of "us" is opposed to "them".

> Politics is not based on an ideology decided in advance, but it is rather constituted through a specific type of narrative that is often called conspiracy theory. This type of theory is always a machination, that is, a narrative mechanism that secretes, as it were, ideological labels such as the right or the left.
>
> (Kelman, 2012, p. 8)

Once politics is reproduced citizens are subject to experiencing threatening events. Fear plays a crucial role in conspiracy-related policies. Unless otherwise solved, modern politics consolidate only where the official discourse is undermined and therefore replaced by another parallel voice. Since conspiracy works from imagination, it needs to fill a gap with allegories, story-telling and fictions. This hidden figure of politics is not only constructed by the secrecy of a complot, or a story that never could be verified, but interpellates the official allegory. By means of creating instability, conspiracy theory is used once and once again to enthrone the elite in the power. Although Kelman's explanation is eloquent and self-explanatory, two major assumptions should be made. The first and most important was the populist character adopted by the US after the attacks of 9/11. Second, Kelman is wrong with respect to the reasons for his diagnosis. Conspiracy elements work jointly with ideology, revitalising social trust whenever ideology does not suffice to exert influence over the citizenry. In an ever-changing world where information is being produced cyclically and distributed to all classes, something that challenges the authority of elites, ideology is not enough to convince sceptical international public opinion. In the moment ideology fails, conspiracy plays a role by restoring the trust in the efficiency of social institutions to protect citizens' rights.

The allegories produced by Kirchnerism to indoctrinate its militancy allude to a much deeper mythical hell: the crisis located in 2001 is where they come from. Counterfeit politics are the result when "the narcissist image" of omnipotence is certainly replaced by the principle of reality. Neither left- nor right-wing governments in Argentina were efficient in administering the economy in a sustainable way that prevents social instability. Because Argentina is still a poor country that occupies a marginal role in the international division of labour, the myth of a country "rich in culture and natural resources", which was historically disposed by some powerful nations such as the United States and the United Kingdom, persisted. The introduction of a plot to rival against "emptied enemies" can only be developed by undermining the credibility of journalism as a social institution, while in doing so Kirchnerites take advantage of an old tension between journalists and academicians.

Kirchnerism and Macrism: two sides of the same coin

To some extent, one might speculate that populism consists in the introduction of religiosity into politics. At the least, this is the point the present section demonstrates. Kirchnerism and Macrism show many of the shared elements populism conserves such as: the diaspora towards a new land (the mythical need of foundation) and the sacralisation of the dead and the disappearance of their bodies, upon which all religions are based. Let's frame the context from where Nestor Kirchner accesses the presidency of the country, as well as the most relevant historical events that accompanied him. Kirchnerism spans a period from 2004 to 2015, when Daniel Scioli was defeated in free elections by the "Alianza-Cambiemos" led by Mauricio Macri (the son of Franco Macri, a well-known contractor). Macri launched himself into politics by taking advantage of the tragedy of Cromañon where 194 persons died. This nightclub (Republica de Cromañon) had a legal authorisation to operate but there were many shortcomings regarding controls over its maintenance. As a result of this, the former Major Anibal Ibarra was tried and deposed after an "impeachment" encouraged by the Pro and Mauricio Macri. This founding event started a tension between "Frente para la Victoria", the party that made Nestor Kirchner president and Pro, Mauricio Macri's political project. The narrative of both sides appealed to "an original hell", a nightmare to which Argentinians should never return. The Kirchners and Kirchnerites alluded to the stock and market crisis of 2001 as the beginning of a diaspora, a new foundational event that called for the glory of the country. Neoliberalism generally and the neo-populism of Menemism not only wreaked havoc in the economy, but also led the nation to one of the worst economic crises in its history. With an unemployment rate of 40 per cent, the financial system almost in bankruptcy and some cryptocurrencies circulating elsewhere, Fernando de la Rua's government suddenly ended on 21 December 2001. Per the Kirchnerist liturgy, neoliberalism failed in promoting development as well as a stable political system in the region (echoing the ideals of Carlos Saul Menem). Considered as one of the most corrupt presidents in history, Menem was demonised as the one responsible for all Argentinian evils. Mauricio Macri arrived in politics supported by Carlos Menem, while his father Franco Macri was in collaboration with the Juntas in order to secure further contracts.

Kirchnerism represented a re-foundational event that hypothetically placed Argentina on the correct side. In the Kirchernites' eyes, the presidency of his wife, Cristina Fernandez de Kirchner, was characterised by a secret alliance with some "rogue states" such as Venezuela and Iran, while she confronted the US, the UK and other European powers. In this respect, Cristina Kirchner's administration was defeated in elections after almost a decade, when some serious economic problems surfaced. The crisis happened in 2008 and ushered Cristina Kirchner into some bad decisions that allowed Mauricio Macri to win the election in 2015. Inflation and political corruption were two important factors Macri capitalised upon in his campaign. Under the slogan *"se robaron todo"* (they ransacked everything), the anti-Kirchernites – accompanied by Clarin and its monopolies – gradually undermined her credibility before the lay-citizenship. Macri not only swore to lead Argentina out of the crisis that Cristina had left the country in, but also offered a new "diaspora", now that the nightmare was Kirchnerism and populism. Though Mauricio Macri shrugged off populism he followed the same political tactics as his predecessors.

The second and third elements both groups shared was "the sacralisation of the dead" and the sentiment of victimisation that opens the door to "the disappearance of the founding hero". For Kirchnerism, the liberal policies, which started just after Maria Martinez de Peron was overthrown by the Juntas, were accompanied by a bloody repression that kidnapped illegally and killed almost 30,000 innocent citizens. These dissidents known as *"los desaparecidos"* (the disappeared dissidents) illustrated one of the saddest years of Argentina. The defeat in the Malvinas/Falklands war stopped the Juntas' hegemony creating a climate of instability that ended in the return of democracy in 1982. Although, Raul Alfonsin paved the way for the trials of the commanders as well as the leaders of the guerrillas, Carlos Menem gave a pardon to all of them. Kirchnerism ignited a fierce debate revolving around the horrendous crimes perpetrated by the Juntas and facilitated the conditions for new trials that indiscriminately involved top-ranking and low-ranked officials who had committed torture and human rights violations.

Anthropologically speaking, religions are often based not only on the disappearance of the "heroes' bodies". Christ, Mohammed or Buddha were never found by archaeologists, in which case this point suggests that religion encompasses the mystery of the body. The mythical hero often is subject to countless obstacles, tribulations and problems, which are posed by the gods as a test of human character. Most certainly, heroes are not only outstanding, they suffer in order to sublimate into a sacrifice for the sake of humanity. They give their lives for others to live, and hence they never die; they live forever in the memory. Whenever a community is obliterated after a disaster strikes, survivors realise that after all the destruction not all is lost. They have survived, touched or protected by God. This mythical narrative starts a process of healing which is needed for the society to recover in post-disaster contexts. However, unless this sentiment of superiority is mitigated, the self may go through a sentiment of radical patriotism or chauvinism that affects the contact with alterity. One of the aspects that define survivors is the belief they have survived for some goal. In that way, they certainly sublimate the tragic loss for a mythical mandate.

Los desaparecidos fulfil a similar role, creating a "floating signifier" which is created by Kirchnerism to impose policies that otherwise would be neglected. Kirchnerites vindicate "the crimes of the Juntas" as acts of evil, while they present their policies as the only option for progress and national liberation. For Kirchnerism and Kirchnerites, *los desaparecidos* should be re-memorised as "heroes" who fought against the liberal forces and the Juntas' liberal policies, while the disappearance of their bodies ignited a new cult oriented towards giving people back upward social mobility and realising the ideals of Juan Domingo Peron and Eva Duarte de Peron. These mythical narratives are the property of Kirchnerism but, as we shall see next, are also shared by Macrism.

In a similar way, Mauricio Macri remembered when he was illegally detained by criminals for a number of weeks. In 1991, Macri was kidnapped by a "ring of top-ranking officers" that formed part of the Argentine Federal Police. While he was placed in a small room, deprived of food and water, he was frightened for his life. Once his father Franco Macri paid the ransom, he was released. This traumatic event not only led Macri to serious problems with sleeping over the years, but also inspired him to enter into politics. He, like Moses, had a dream, a vision given by the gods. First, he started with Boca Juniors and once the sporting triumphs endorsed his presidency he presented himself as a candidate to succeed Jorge Telerman, who had replaced Anibal Ibarra after the above-noted impeachment. Just like Cristina and Nestor who supposedly were persecuted by "the military forces" (*los milicos*), Macri was a victim of power, the police whose corruption had no limits. Macri and Cristina are presented as mythical heroes, though while the former goes on to his own individuality – as the chosen leader who suffered – Cristina appealed to the collective memory of the NRP (National Reorganization Process). Kirchnerism imagined a better destiny for the country leading Argentinians from the nightmare of 2001, and Macrism does the same, by liberating the economy from Kirchnerism. Bascially, as stated, both policies followed the same neoliberal dynamics that systematically combined what Kelman dubbed "counterfeit politics" with higher levels of populism. The figure of evil, last but not least, occupies a central position in the populist doctrine. Here two assumptions should be adopted. On the one hand, the presence of a hero needs the controversial anti-hero. The archetype of evil in Kirchnerism seems to be the "neoliberal discourse", the "free market", as well as the International Monetary Fund and its miscarried policies that finally ushered Argentina into a terminal crisis, and of course the looming forces of Mauricio Macri who had been complicit with the Juntas. Military forces, in sharp rivalry with the popular movement, not only proscribed Peronism in the past, but also eradicated a generation of intellectuals, writers and dissidents who bravely opposed the advance of neoliberal order. The ruling elite, protected by the Clarin Corporation, worked hard to dismantle the national industry while adopting decentralised forms of consumption that ruined the economy. Like the Devil tricking humanity to fall into sin, Clarin distorts reality in order to protect what Cristina Kirchner dubbed "the interests of monopolies" or the "agrarian aristocracy". The mass media, and journalism in particular, played an active role in harming the credibility of the government. She acknowledges that (following Laclau) the sense of truth is individually constructed and relative to what the media draws. For this reason, *"el periodismo militante"* – militant journalists – are not only necessary but a priority for a national and popular government. For the Kirchners, militancy and the appropriation of public space, not the virtuality of the media, would be the cure for the anti-politics that Mauricio Macri offers. Nestor Kirchner, who suddenly died during the first presidency of Cristina, gave his life for the nation, or so Kirchnerites think. His body was never shown in public to the lay people who congregated to pay their last respects to his remains, which surely emulates the ascendancy of Christ, Mohammed and any other mythical heroes.

On the other hand, Macri sees the evil in Kirchnerism, political corruption and the various overcharges generated by the Kirchners over more than a decade in the presidency. Macri knows he is a select person, an envoy of God, who survived the terrible conditions he went through when he was detained, but he says "I am good at the process of team formation". Of course, this led him to claim overtly that his team is the best in Argentina's history. Macri vindicates the helpful support of the IMF and valorises the benefits of free trade. With the benefits of hindsight, he is strongly convinced Argentina has a great future, because Argentinians are stronger, smarter and very innovative, but he sees in Kirchnerism and workers' unions the major obstacle to "genuine economic growth". He is receptive to the critiques of journalists, but he sees the world cynically in a one-sided way. Cristina appealed to the past times, whereas Macri believes in an idealised future and yet erroneously bets on a world which today no longer exists in the wake of the 2008 stock-market crisis. The salvation of the local economy is the flurry of foreign investments that are possible because of the transparency and credibility of his administration. For his viewpoint, Kirchnerites are partisan, egoists, irrational, evildoers, if not filled with hatred and a potential threat to be controlled. Kirchnerites are part of what he calls "the irrational Peronism" – which is contrasted with "the responsible Peronism" – and is a reminder of the dangers of a return to Kirchnerism. They are involved in the resentment and pain that left the "dirty wars", while he is the hope of a new, renovated and united nation. Cristina divided to rule, while he is engaged in a struggle to keep Argentinians united, as a beloved family.

Doubtless, fear is a centrepiece of populist leaders and Kirchnerism and Macrism show why – as the title of this chapter indicates – they are two side of the same coin. In a nutshell, populism and neoliberalism are inextricably intertwined.

Conclusion

The present chapter has discussed the myths derived from populism, which in Argentina shaped two opposing forces. While Kirchnerism appealed to a dramatic liturgy, which is enrooted in the political violence of the 1970s, Macrism elaborated a mythical discourse with foci on an idealised future that calls upon Argentina to be a great country once again. Emulating Obama's slogan "yes we can!", Macri founded a new political coalition that replicated the conspiracy as the centrepiece of politics. Like Kirchnerism, Macrism continued with the neoliberal narratives and factional democracy. As Kelman observed, "the counterfeit politics" are characterised by the introduction of mystery as the centre of politics. At the time an official history imposes, there is an alternative plot, which is emptied and filled according to the interests of the ruling elite. This chapter has interrogated what journalism adopted as *"la grieta"* (the crack), but far from seeing in this a point of conflict, discrepancy and disunion, it defines *"la grieta"* as an ideological platform to make lay-citizens believe that they are part of something important. Like the Saga of the Matrix, where humans are enslaved and connected to pods in order to serve as sources of energy alone, while they live in a great simulacrum, populism orchestrates "a great fable" combining "narcissism" and the mythological elements of religiosity in order for the ruling class to posit policies that otherwise would be neglected.

Note

1 After being elected, Nestor Kirchner entered office as president on 25 May 2003 to be replaced by his wife, Cristina Fernandez de Kirchner, in two periods (2007–2011 and 2011–2015). Her candidate for president, Daniel Scioli, was defeated by the current president, Mauricio Macri, who came into office on 10 December 2015.

References

Arendt, H. (1964) *Eichmann in Jerusalem*. New York: Viking Press.

Arendt, H. (1972) *Crises of the Republic: Lying in Politics, Civil Disobedience on Violence, Thoughts on Politics, and Revolution*. New York: Houghton Mifflin Harcourt.

Arendt, H. (2013) *The Human Condition*. Chicago: University of Chicago Press.

Armitage, J. (2011) *Virilio Now: Current Perspectives in Virilio Studies*. Cambridge: Polity Press.

Baudrillard, J. (1985) 'The masses: The Implosion of the Social in the Media', *New Literary History*, 1, pp. 577–589.

Baudrillard, J. (2006) 'Virtuality and Events: The Hell of Power', *Baudrillard Studies*, 3(2), pp. 1–18

Bird, E. (2010) 'Introduction', in *The Anthropology of News & Journalism: Global Perspectives*. Bloomington: Indiana University Press, pp. 1–20

Blumler, J. (2014) 'Mediatization and democracy', in F. Esser and J. Strömbäck (eds) *Mediatization of Politics: Understanding the Transformation of Western Democracies*. New York: Palgrave Macmillan, pp. 31–41

De Vreese, C. (2014) 'Mediatization of News: The Role of Journalistic Framing', in F. Esser and J. Strömbäck (eds) *Mediatization of Politics: Understanding the Transformation of Western Democracies*. New York: Palgrave Macmillan, pp. 137–155.

Esser, F. and Stromback, J. (2014) 'Mediatization of Politics: Towards a Theoretical Framework', in F. Esser and J. Strömbäck (eds) *Mediatization of Politics: Understanding the Transformation of Western Democracies*. New York: Palgrave Macmillan, pp. 3–30

Frankfurt, H. (1987) 'Equality as a Moral Ideal', *Ethics*, 98(1), pp. 21–43.

Grossman, M. (2014) *Artists of the Possible: Governing Network and America Policy*. Oxford: Oxford University Press.

Ignatieff, M. (2003) *Human Rights as Politics and Idolatry*. Princeton: Princeton University Press.

Kelman, D. (2012) *Counterfeit Politics: Secret Plots and Conspiracy Narratives in the Americas*. Lewis, PA: Bucknell University Press; Lanham, MD: Rowman & Littlefield.

Korstanje, M. E. (2014). 'El miedo político bajo el prisma de Hannah Arendt', *Revista SAAP: Sociedad Argentina de Análisis Político*, 8(1), pp. 99–126.

Korstanje, M. E. (2016) 'Tergiversation of Human Rights, Deciphering the Core of Kirchnerismo', *International Journal of Humanities and Social Science Research*, 2, pp. 60–67.

Lang, B. (1990) *Act and Idea in the Nazi Genocide*. Chicago: University of Chicago Press.

Lang, B. (1996). *Heidegger's Silence*. Ithaca: Cornell University Press.

Lara, M. P. (2007) *Narrating Evil: A Postmetaphysical Theory of Reflective Judgment*. New York: Columbia University Press.

Lasch, C. (1991) *The culture of narcissism: American Life in an Age of Diminishing Expectations*. New York: W.W. Norton & Company.

Levinas, E. (2001). *Is it Righteous to Be? Interviews with Emmanuel Levinas*. Stanford: Stanford University Press.

Lewis, P. H. (2002). *Guerrillas and Generals: the "Dirty War" in Argentina*. Westport: Greenwood Publishing Group.

Mazzoleni, G. (2014) 'Mediatisation and Political Populism', in F. Esser and J. Strömbäck (eds) *Mediatisation of Politics: Understanding the Transformation of Western Democracies*. New York: Palgrave Macmillan, pp. 42–56

Morgan, M. L. (2001). *A Holocaust Reader: Responses to the Nazi Extermination*. Oxford: Oxford University Press.

Peperzak, A. (2013) *Ethics as First Philosophy: The Significance of Emmanuel Levinas for Philosophy, Literature and Religion*. Abingdon: Routledge.

Simon, J. (2007) *Governing through Crime: How the War on Crime Transformed American Democracy and Created a Culture of fear*. Oxford: Oxford University Press.

Virilio, P. (1995) *The Art of the Motor*. Minneapolis: University of Minnesota Press.

Virilio, P. (2005) *City of Panic*. London: Berg.

Virilio, P. (2010) *The University of Disaster*. Oxford: Polity Press

Žižek, S. (2002) *Welcome to the Desert of the Real! Five Essays on September 11 and Related Dates*. London: Verso.

Žižek, S. (2004) 'A Plea for Ethical Violence', *Bible and Critical Theory*, 1(1), pp. 1–15.

Žižek, S. (2005) 'Against Human Rights', in A. Singh Rathore and A. Cistelecan (eds) *Wronging Rights? Philosophical Challenges to Human Rights*. New Delhi: Routledge, pp. 149–167

Žižek, S. (2015) *The Universal Exemption*. London: Bloomsbury.

4 Vox of whom?

An assessment of Vox through discourse analysis and study of the profile of its social base

Alejandro Pizzi and Verònica Gisbert-Gracia

Introduction

In many Western countries, and in particular in Europe, there is a crisis of representation and a political vacuum (Castells and Pradera, 2018), apparent at least since the first decade of the twenty-first century. Some societies were closely linked to the global mobilisation cycle of 2011, which occurred in specific countries (Occupy Wall Street, 15M in Spain, Greece, etc.). Other societies had less intense connections, more ambiguous or even non-existent with these processes of mobilisation. From the first decade of the twenty-first century, in Europe there has been an unequal distribution of social mobilisation and the intensity of the emotions involved in these processes. In this context, various conservative/popular political formations have taken advantage of this political gap in certain countries to grow electorally and influence their political cultures more directly. In this regard, some political and intellectual observers claim that Europe is experiencing a "populist moment" of a reactionary nature (Mouffe, 2018). This is because they perceive a growing separation or disconnection between the elites and the wider populace, who lack adequate representation. Others interpret the phenomenon in terms of the expansion of "conservative populisms" or "right-wing populisms", given that the citizens' feelings about the crisis creates favourable conditions for the reception of proposals for strengthening the authority and power of national states, together with the social exclusion of all identities that are perceived as threatening their own culture. These threats range from immigrants, below, to the institutions of the European Union, above, which appropriates areas of sovereignty of the old national states. In this regard, there is an academic current that discusses the relationship between populism and liberal democracy (Canovan, 2005; and Rovira, 2017; Dassonville and Hooghe, 2018).

The structural conditions that favour the acceptance of these reactionary discourses are linked, on the one hand, with the post-Fordist transformations of the European economy and culture over recent decades (Thompson, 2003, 2013; Fumagalli, 2010; Fumagalli, 2011). Due to this, there are entire economic sectors in different countries of Europe that have lost competitiveness, perhaps permanently. On the other hand, demographic-structural trends of population aging are recorded (Eurostat, 2018) that affect European auto-chthonous social identities. These two processes occur simultaneously over time and reinforce feelings of uncertainty, anguish and fear of the present and the future. Other approaches, rather than interpreting the present in a populist way, prioritise analysing the political regimes of labour control that are being tested in different states (Lorey, 2016). Here different versions of the "moral state" appear. In these arenas, we have the post-colonial dimension present in the "central" states (Galcerán, 2016; Mezzadra et al. 2008), one of whose most visible aspects is the growing number of poor immigrant workers, pre-cariously inserted into European labour markets, and that the systems of social security need to be financed. This constitutes an expression that the "colony" is already present in the interior of the metropolis. At the same time, we are witnessing capitalist transformations that modify the forms of work and labour relations (Visser, 2013; Marginson, 2015). The two trends (pro-ductive and demographic changes) are linked to new political regimes of labour control. Considering that ethnic homogeneity is no longer possible in European societies, the new modes of labour control seek to constitutionalise social differences in terms of a certain "apartheid", while promoting a regime of population control that needs immigrants to work in conditions of max-imum exploitation with lower social, labour and political rights. It is a model that has features of "apartheid" because it needs to control and govern the mobility of a differentiated segment of people within a territory.

The preceding paragraphs are intended to synthetically describe the context in which the socio-political dynamic in Spain is inserted.

In particular, the mobilisation cycle of 15M from 2011 implied some changes in important aspects of the frames of the dominant political culture. This resul-ted in the emergence and growth of political parties, social movements, plat-forms and new confluences that organised the system of political senses and emotions in a democratic key. However, in recent years political sensitivities in Spain are undergoing new changes in intensity and content. The electoral growth of the radical-right formation Vox indicates that its discourse includes social demands to reinforce the internal social order (understood in a conservative key), as well as collective aspirations to strengthen the sovereignty of the Spanish state against everything that they consider "threats" to their unity.

In this sense, it seems to be aligned with political movements in other European countries that present similar approaches in the same continental context. To what extent does this political formation address a collective sensibility that is on the same frequency as the so-called "conservative populist" or "neo-fascist" movements of other European countries? In this regard, the first objective of our contribution is to analyse the central ele-ments of Vox's political discourse, with the aim of studying its narrative keys to interpret how it articulates with the rest of the conservative populist political forces (or neo-fascist) in Europe. The second objective is to statis-tically describe the political-ideological and socio-demographic profile of citizens who feel an affinity towards Vox's approaches.

In this way, we present a picture that integrates the analysis of the central elements of the discursive framework of this political movement, on the one hand, and the analysis of the perceptions, opinions and social features of those who feel affinities towards Vox on the other. All this offers us a panorama to evaluate the articulation of this Spanish socio-political force with the wave of reactionary and xenophobic movements that Europe is going through.

Discursive analysis of Vox

In this first section of the chapter we offer a synthetic analysis of the main elements that make up Vox's discourse. We base this analysis on its electoral program, as well as its founding manifesto. From these documents we have obtained the main signifiers and meanings that organise the central elements of their discursive frame. Based on this, we elaborate an analysis that allows us to visualise the main aspects of its political position. This elaboration allows us to evaluate the political logic of their approach, and the connection that it has (or not) with the populist and/or authoritarian logics. We can also investigate its connections with the collective emotions and impulses promoted by the forms of neoliberal governmentality that have been dominant since the end of the twentieth century and the beginning of the twenty-first century.

The greatest and most important value of the political order in Spain for the political party Vox is the sovereign power of the Spanish state. From this point of view, it guarantees national unity and territorial integrity as a constitutional monarchy. Thus, according to this discourse, in order for the Spanish state to function well in political terms, but also to constitute the base of the economic growth, it must guarantee the existence of a territory and a unified and disciplined population founded on traditional values, and manage in a homogeneous way its different regions. In relation to the identity of the subject that must sustain the political order, the structural place of the "us" Vox aspires to is made up of individuals who identify with the Spanish nation absorbed in values of patriotism, meritocracy and traditionalism (defence of the traditional family, prohibition of abortion, promotion of the Catholic religion, etc.). Moreover, like any identity, it is differentially established and defined by the opposition relationship with other identities that threaten it. These threatening identities ("others") are made up of the following features: 1) "peripheral nationalisms" that challenge territorial unity (in the Spanish case, the Catalan independence movement); 2) the immigration that threatens the cultural traditions and the resources and economic opportunities of the Spanish people; and 3) socio-political movements that emphasise the violent origin of the current political regime (as well as the discursive elaboration of an alternative Spain with some republican and social contents) synthesised in the impulse and defence of the laws of historical memory.

On the one hand, the unity of the country (guaranteed by the strength of the central state) and, on the other, the traditionalist identity of the Spanish people constitute the central most determining categories of Vox's political discourse, and are structurally linked such that they require each other to sustain the discourse. Both are defined in relational terms, each one based on its relations with "the other" that defines them as such.

Next, we describe the strongest elements of Vox's discourse that positivise these differences. Vox defends the project of a unitary state. The autonomic[1] powers did not achieve their initial objectives, which were to appease the centrifugal territorial tensions of the Spanish state. On the contrary, according to their perspective, they have expanded them and promoted their territorial and political disintegration. In this sense, we can say that for Vox the local autonomic powers occupy, in the structure of the story, the conventional place of the oligarchies that, in the typical populist discourse, threaten the integrity of the people (in this case, Spanish). These local territorial powers are seen as oligarchies because, according to Vox, they constitute political groups or structures that have captured the state and are sustained materially thanks to the continuous extraction that they carry out from public resources, with the objective of reproducing themselves as such and benefiting only themselves, to the detriment of the Spanish people.

The increase of the territorial and political tensions on the part of the independence movements have put at risk the sovereign power of the state in jeopardy (the maximum value of the political order), for which the organisations and subjects that promote it are classified and treated as "enemies". Especially, in the context of recent years, the logic of friend/enemy applies to the Catalan independence and sovereignty movement, but it is also extended to the Basque and Navarrese movements, and to all those who carry out policies that question the territorial and cultural unity of the central state. As a result of this, the solution to this challenge is not the integration of diversity, but the defeat and submission of these identities and alternative interests, and this translates into a state centralisation and a unitary political regime.

Moreover, the Spanish identity is also threatened by the figure of the "immigrant". In this case, the attack focuses on their cultural and religious traditions, as well as on the economic conditions of life due to the pressure exerted by foreigners on the welfare state and on the job market. Particularly stigmatised are the poor immigrants from countries of Africa and the Muslim religion, but also those from Latin America. In this regard, as in speeches by other movements known as populist-conservative or "neo-fascist", the figure and signifier of the "wall" that should be constructed or reinforced to protect the territory from the migratory threat on its own culture is prominent. In this way, the discourse of Spanish identity is constructed through the game of differences with other threatening identities, in the face of which the only alternative is a strategy of confrontation and exclusion.

In this discursive structure, the place of "democracy" is hierarchically subordinated to national unity and territorial integrity. Both constitute a limit to the exercise of democratic rights. In this sense, we can say that behind this limit is the figure of the enemy, not that of the political adversary. For this reason, Vox's speech excludes from the democratic space all those political formations and social organisations that promote internal independence processes. In connection with the economic model, Vox's discourse differs in part from other variants of the aforementioned conservative populisms that propose forms of state control and regulation of markets, with the aim of protecting the people from the direct effects of globalisation. Nevertheless, Vox proposes an open and deregulated economic model, based on the guarantee of property rights and free entrepreneurial initiative. This model, for Vox, requires that the population be socialised with strong moral values, which have proven to endure over time, and are related to patriotism and national traditions.

These values are what will give willpower to the Spanish people to sustain this project in the long term. Specifically, Vox's proposal suggests an economic model based on income transfer in favour of the business sectors, through tax reductions (reduction of tax pressure on companies, moderation of social contributions of "Spanish" private and public salaried workers, elimination of added taxes on utility bills, etc.), together with cuts in public spending (elimination of regional state structures that duplicate functions and costs, elimination of public subsidies to political parties, unions, business associations, etc., promotion of private education, incentives for private pension schemes and insurance based on individual savings, etc.). This approach is complemented by a strong state interventionism in the promotion of nationalist, traditionalist and socially disciplining values.

Finally, the link with the EU is subordinated to reserving the resources of power to defend the territorial sovereignty of the Spanish state. We can see that, in the structure of the narrative, both democracy and insertion into Europe occupy a secondary position with respect to state centrality. In this way, the structural articulation of all the elements analysed suggests that Vox's discourse does not specifically adjust to the positive contents of conservative populisms at the beginning of the twenty-first century. On the contrary, it seems to us that it is closer to liberal political logics of state organisation, in line with experiences of Bolsonaro's Brazil, Orbán's Hungary, Putin's Russia, and so on.

In Figure 4.1, we synthesise this analysis in a semantic network that highlights the main discursive elements and their structural position within Vox's narrative.

The discursive keys that appear marked in the central part of the semantic network (the "Spanish state", the "Spanish identity", the "unitary state", the "traditional values" and the "repression of threats") constitute the main axes (but it does not exhaust it) of the speech of Vox, that revolving around the centralised state as supreme value of the social and political order, a homogeneous Spanish identity, and the necessary policy to guarantee said central axis: on the one hand, the cultural promotion of a traditional and patriotic morality, and, on the other, the criminal repression of those who are located outside the limits of the project of the unity of Spain.

The main aspects that threaten the project of unity and sovereignty of the Spanish state are the "autonomic oligarchies" that exercise centrifugal forces and deplete public resources inefficiently, the independence movements,

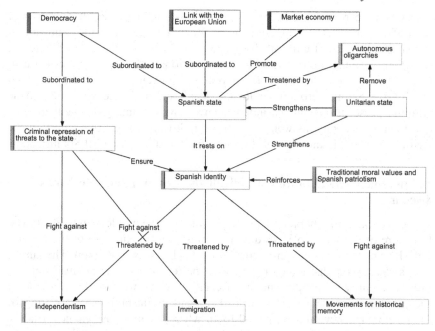

Figure 4.1 Semantic network of the discursive structure of Vox
Source: own elaboration (2018)

immigration and projects offering a different model of Spain (which are syn-thesised in the defence of historical memory).

In turn, "democracy" and "integration to the EU" constitute the discursive aspects that are subordinated to the central axis of the authoritarian model of Vox. Clearly, they constitute secondary elements of the discursive structure of the political formation that we are analysing because its limits are defined by the effectuation or not of the sovereignty and unity of the Spanish state. Finally, we would like to mention that the proposal to strengthen a deregulated market economy reveals that we are not facing attempts to regulate the effects of globa-lisation (except for the foreign labour force), but to fully insert ourselves into it.

Thus, our synthetic interpretation is that we are facing a project of the dominant classes linked to financial globalisation (which demands flexibility of the different markets, especially the job market) and central state bureau-cracies (which demands the concentration of decision-making power within the territory), and which is based on creating conditions of economic dereg-ulation in favour of the free market, but also population and political reg-ulation in a conservative key. As a result, its objective is to discipline the population through the socialisation of traditional and hierarchical values, as well as to exclude from the space of legitimate politics the forces that propose alternatives to this domination scheme with conservative features.

For Vox, therefore, the necessary condition to achieve the liberalisation of the different markets throughout the territory, on the one hand, and homogenising/political and cultural discipline exercised from the central state, on the other, is to tear down the regional institutional differences and centrifugal forces that are held in them; defeat political movements defending a social and plural Spain; through territorial, legal, cultural reforms; and pursue criminal prosecutions against independence forces claiming the right to exercise their sovereignty, as well as all those political projects that question the legitimacy of the foundation of the centralised and monarchical Spanish state.

Political-ideological profile of individuals with an affinity to Vox's discourse

In this section of the chapter, we propose to analyse the profile of individuals who feel more or less close to Vox's discursive positions, making use of the January 2019 Barometer from the Centre for Sociological Research of Spain. The sample size is 2989 cases, distributed throughout the territory of the Spanish state. In order to identify the people who may feel close to Vox, we consider a variable of the questionnaire, using a Likert scale, that measures the probability of voting for the proposed policy on a scale of 0 to 10 where 0 equals "no chance" and 10 represents the "most likely" to vote for Vox in the Spanish general elections to be held on 28 April 2019. From this variable, we construct new data grouped into two categories: "Close to Vox" (includes all those who were assigned a probability of voting Vox between 5 and 10 points), and "not close to Vox" (which includes those who were assigned a chance of voting for Vox between 0 and 4 points). Thus, we have a simple dichotomous variable that simplifies the object of our analysis, as we use it to construct an approximate knowledge of the profile of people who feel some degree of proximity to the political formation of Vox. Also, when we find it useful to specify a little more data, we develop a variable measuring "Vox affinity level" (low level, high level, and intermediate) through the distribution into three groups of original scale of 0 to 10.

Once we have developed these variables, we relate them to a set of other variables of a political-ideological and demographic nature that are part of the barometer, in order to explore the features presented by the individuals more or less close to Vox. If we start from the hypothesis that Spanish nationalist feelings favour the affinity towards Vox, we see that among those who have a high Spanish nationalist sentiment, the proportion of Vox-like people is much greater than the proportion of Vox-like who have low Spanish nationalist sentiment (a proportion ten times higher, 23 per cent compared with 2.3 per cent). We see a direct relationship between Spanish nationalism and affinity with Vox. Of course, many other individuals who also consider themselves Spanish nationalists have more affinities with other parties, even those opposed to Vox (like the PSOE). However, that does not contradict the clear relationship we have detected and discussed above; among those who have low Spanish nationalist feelings there are almost no individuals with affinities towards Vox (2 per cent).[2]

Table 4.1 Proximity to Vox, according to Spanish nationalist feelings

| | | | Vox's Proximity | | Total |
			Not close	Close	
Spanish nationalist feelings	Low	Count	469	11	480
		%	97.7%	2.3%	100%
	Medium	Count	1375	181	1556
		%	88.4%	11.6%	100%
	High	Count	403	120	523
		%	77.1%	22.9%	100%
Total		Count	2247	312	2559
		%	87.8%	12.2%x	100%

Source: personal compilation-based CIS Barometer January 2019

The way the nationalist and xenophobic discourse of Vox sets and shapes the political emotions explains the fear and insecurity about foreigners this speech causes among like-minded individuals. This can be seen in Table 4.2, in which we see what are, for those who feel close to Vox, the main problems facing Spain. Specifically, we analyse the weight of those problems that imply challenges to the unity and homogeneity of Spain. So, we see that, among those who feel favourably towards Vox, the "challenges to the unity of Spain" (which refer especially to the Catalan challenge and the aspirations of independence of the Basque Country) and the problems related to "immigration and refugees" have almost three times the percentage frequencies other problems facing Spain (27 per cent and 28 per cent respectively, compared with 10 per cent for other problems). In this sense, the discursive elements that Vox frames as threatening the unity of Spain attain three times more recognition than other problems among those who are oriented towards this political party. The observed differences are statistically significant percentages.[3]

If we also look at the personal problems that most concern those who have affinities with Vox's discourse, as recorded in Table 4.2, it is clearly highlighted that the problems related to immigration and refugees, insecurity and terrorism, and challenges to the unity of Spain (which is represented by the Catalan independence movement) have a double and triple proportional weight (33 per cent, 25 per cent and 20 per cent respectively) with respect to other problems (between 6 per cent and 12 per cent). Thus, among those who have affinities with Vox, we note that the problems they considered general for the whole of Spain are also the same as those they see as their own personal problems (at a higher rate than other types of problems). Here also appears greater insecurity and terrorism, which also refers to fears arising from xenophobic discourse, promoted as collective and politically manipulated emotions.

Table 4.2 Main problems of Spain, according to proximity to Vox

| | | | Proximity to Vox | | Total |
			Not close	Close	
Main problems of Spain (regrouped)	Economic problems	Count	1064	119	1183
		%	89.9%	10.1%	100%
	Crisis of welfare state and social security	Count	126	14	140
		%	90%	10%	100%
	Insecurity and terrorism	Count	23	3	26
		%	88.5%	11.5%	100%
	Challenges to the unity of Spain	Count	67	25	92
		%	72.8%	27.2%	100%
	Immigration and refugees	Count	53	21	74
		%	71.6%	28.4%	100%
	Corruption and lack of moral values	Count	307	37	344
		%	89.2%	10.8%	100%
	Social discrimination and environmental problems	Count	62	3	65
		%	95.4%	4.6%	100%
Total		Count	1702	222	1924
		%	88.5%	11.5%	100%

Source: personal compilation-based CIS Barometer January 2019

With respect to the ideological profile of individuals who have an affinity with Vox's discourse, the data show that among those who consider themselves conservative, 38 per cent feel affinity for such a political formation. Among those who consider themselves Christian democrats, 23.3 per cent feel in agreement with Vox. Then, among those who identify with other ideologies only a minority of liberals and nationalists (16 per cent and 14 per cent respectively) feel affinity for Vox, while other cases are negligible percentages (progressive, social democrat, socialist, communist, feminist and ecologist). The importance of conservative self-definition (including Christian democracy) over the rest of the ideologies suggests to us that, among the Vox-like individuals, the "plebeian" self-perception of any typically populist movement would be absent. We are, rather, facing a conservative social movement, formed by citizens who prefer the authority of the central state to guarantee the social order rooted in the traditionalist hierarchies of Spanish social history.

The statistical data also indicate that the perception of the challenge to the unity of the Spanish state constitutes a motivation to electorally support a discourse that poses a deep confrontation with what puts at risk the unity of

Table 4.3 Main personal problems, according to proximity to Vox

| | | | Proximity to Vox | | Total |
			Not close	Close	
Main personal problem	Economic problems	Count	963	133	1096
		%	87.9%	12.1%	100%
	Crisis of welfare state and social security	Count	393	44	437
		%	89.9%	10.1%	100%
	Insecurity and terrorism	Count	21	7	28
		%	75%	25%	100%
	Challenges to the unity of Spain	Count	36	9	45
		%	80%	20%	100%
	Immigration and refugees	Count	24	12	36
		%	66.7%	33.3%	100%
	Corruption and lack of moral values	Count	119	12	131
		%	90.8%	9.2%	100%
	Social discrimination and environmental problems	Count	28	2	30
		%	93.3%	6.7%	100%
Total			1584	219	1803
		%	87.9%	12.1%	100%

Source: personal compilation-based CIS Barometer January 2019

the nation (objectified in the central state). Thus, among those who consider that the Catalan situation influences "enough" and "a lot" of their voting decisions, those who are close to Vox practically double and triple the percentage (23 per cent and 12 per cent) as compared with those who think that it will influence little or nothing about their vote and feel, at the same time, affinity to Vox (7 per cent). This data is considered an indicator that affective identification with Spain is built through strong opposition and rejection of political movements that question it.

Now, what is the demand of those who are oriented to Vox regarding the appropriate policy that the central government should implement towards the Catalan independence movement? According to the analysed discourse, data indicate that those who feel a high affinity towards Vox's positions, in their great majority they consider that a "hard hand" is required, which implies managing the conflict through repressive forms based on penal legislation. By contrast, negotiated solutions to the conflict, in their great majority, hold those who have low affinity for the discursive positions expressed by Vox. These facts show us that, among people related to Vox, everything that threatens the unity of Spain should be treated with the logic of "friend/enemy".

This defence of the integrity of the Spanish state through repressive methods corresponds to the discursive structure analysed, according to which the exercise of democratic rights is limited or subordinated to the defence of the unity of the state and the Spanish nation, given that these constitute, by themselves, values superior to democracy. If we analyse the preferences on the territorial organisation of the Spanish state, according to the affinity towards Vox, dates indicates that among those who are close to this party, those who prefer a unitary state and/or one with autonomous communities with fewer powers and competencies represent a much higher percentage (35 per cent and 19.6 per cent respectively) than those who prefer a central state with equal or fewer powers (6 per cent and 3 per cent, respectively).

Next, we jointly articulate the presented variables in order to reconstruct the profile of political sensitivities close to the Vox discourse. We do this through a multiple correspondence analysis, which allows us to develop an integrated structural position in political-ideological terms, of people who feel challenged by Vox's discourse.

In the perceptual map of the analysis of multiple correspondences we see in an aggregate form the political-ideological features of the individuals close to

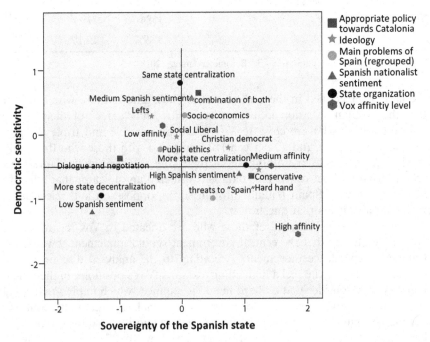

Figure 4.2 Multiple correspondence analysis. Democratic sensibility and the preference for the sovereignty of the state

Source: personal compilation-based CIS Barometer January 2019

Vox. The structural proximity between these features indicates its political profile and can be better visualised by the content we see in each of the quadrants.[4] To visualise the results clearly, we have regrouped and reduced the categories of some variables used. With regard to the main problems of Spain, we regroup them into three general themes. First, the problems for Vox are a "threat to Spain" (the independence movement, immigration and refugees, and terrorism). Second, they concern socioeconomic problems (labour, welfare policies, income, etc.). Third, they relate to the lack of public ethics in political management (corruption, lack of transparency, etc.). Meanwhile, ideologies have been redefined to simplify the number of categories,[5] so as to facilitate the interpretation of the perceptual map, without losing the link information between the great ideological trends and individuals oriented to Vox. The results showing the multiple correspondence analysis allow us to observe the presence of two latent overall dimensions. The first latent dimension, which is expressed on the map horizontally, includes the level of Spanish nationalism and preference for the territorial organisation of the state. We can consider that both variables refer to a single latent dimension that is the strength of the "sovereignty of the Spanish state". Said axis locates on the left the preferences for a lower "sovereignty of the Spanish state" (less state centralisation and less Spanish nationalist sentiment), and on the right the preferences for a greater "sovereignty of the Spanish state" (more state centralisation and more Spanish nationalist feelings). An intermediate level considered keeping the situation as it was agreed in the present constitution. Specifically, in the lower right quadrant we see the structural closeness between the preference for greater centralisation of the state and high Spanish nationalist sentiment, Vox's discourse defined by opposition to anything that threatens national unity. The second latent dimension could be proposed for the other variables and distributed in the vertical axis direction. There are included the ideologies, the perception of what are the main problems of Spain and the most adequate policy of central state to confront the independence movement in Catalonia (considering it an indicator of the appropriate relationship that should be established with the "other"). We take into account that the aspect that links all these variables is what we could call "democratic sensitivity"; that is, the recognition of the other, the ways of approaching and managing conflicts, and the (ideological) ways of conceiving the social order. Generally, at the top of shaft positions are located more "democratic sensitivity", and those at the bottom show a lower "democratic sensitivity". And, more specifically, the upper-left part of the perceptual map shapes the space of those who have low affinity towards Vox. There left-wing and social-liberal ideologies are located in the upper-left part of the perceptual map, as are the negotiated solutions to the conflict with Catalonia, the prioritisation of problems linked to public ethics in state management (and the socioeconomic problems of the country are also very close). All these elements are also related to the preference for a lower level of "sovereignty of the central state" than that which is current, and for greater decentralisation and territorial recognition of the various nationalities that make up Spain. That is, elements that refer to a greater willingness to recognise the figure of the "other" and a willingness to integrate it into the common space of democracy are associated. This space configures a set of structural relationships between sensitivity and affectivities away from Vox.

On the right upper quadrant, we see a group of elements that are structurally related to each other. Predominant preference for an equal and/or greater level of sovereignty of the central Spanish state, but with a greater democratic sensitivity compared with those who manifest a high affinity towards the speech of Vox. In this structural space there is more proportional presence of concern for social problems, a greater tolerance relative to the "other" than the more radical positions (observable in that here it is associated with the option for a combination of "hard hand and negotiation" in relation to who question the unity of the state) and is more associated with Christian democracy. All this quadrant constitutes the politico-ideological ecosystem relatively close positions that have an "intermediate level of affinity with Vox". However, this proximity can be qualified because those who manifest an intermediate affinity are located only at the limit of this quadrant; that is, they are also at the limit of the structural space of high affinities towards Vox. This indicates us that those who have an intermediate affinity towards Vox are much closer to the discursive universe of those who have high affinity than those who have low affinity to Vox.

In this sense, we analyse the characteristics of the quadrant of "high affinity with Vox" (lower right). There we see its structural link with the preference for the greater sovereignty of the state and its lower democratic sensitivity. Specifically, in terms of the desire for greater state sovereignty, they are individuals with a high Spanish nationalist sentiment and with the will to centralise more the power of the state. In relation to their lower democratic sensitivity, people related to Vox consider that the main threats come from "the other" of Spain (immigrants, refugees, separatists, terrorists), which is in line with a speech from intolerance and lack of solutions that integrate political and identity differences in a common narrative. For this reason, they support repressive solutions to the Catalan separatist conflict of challenge to the state (the "hard hand"). These fears and drives are consistent with the discourse of exclusion and expulsion of immigrants and refugees that threaten the "Spanish" culture of their own. Finally, the ideology most associated with this sensitivity is the conservative one. That is, we are facing a structural space that presents traditionalist ways of conceiving the social order and rejects the recognition of everything considered "other" of the Spanish nation, through repressive strategies.[6]

For these reasons, the analysis of multiple correspondences allows us to see that people related to Vox's discourse prefer a greater sovereignty of the Spanish state and manifest a lower democratic sensitivity. All this is consistent with a sensitivity related to the discourse structure that we have described in the first part of the analysis, in which democracy is subordinated to the defence of state sovereignty, and Spanish identity is defined by opposition to threats that come from what is defined as the "other", against which fits a strategy of confrontation and repression.

Socio-demographic profile of individuals related to Vox's discourse

In this section, we analyse the socio-demographic profile of people oriented favourably to Vox's speech. With regard to the age of the individuals who feel drawn to Vox's discourse, we see that the proportion of support is greater among the mature-young population, as well as older (among 21 per cent and 23 per cent), and is lower among youth (among 6 per cent and 12 per cent). Specifically, the age groups that proportionately more support these narratives are those between 35 and 54 years and those over 65 years.

Regarding sex, we see that there is a higher percentage of men (58 per cent) than women (42 per cent) who feel close to Vox's speech. In part, this could be explained by the strong presence of traditionalist and conservative discursive contents, especially a set of issues related to the defence of patriarchy as a central institution of the social order posited by Vox's proposals.

With regard to the relationship between the type of occupation or professional situation and the affinity to Vox, on the one hand, we observe its weight among wage earners, self-employed workers and entrepreneurs. However, to determine if there is any specificity in the case of Vox, we must compare it with those who do not feel this affinity. To do this, we cross-tabulated, and we see that this percentage differs according to those who are salaried, self-employed and entrepreneurs, and the extent to which they have (or do not have) an affinity is statistically significant[7] and deserves comment. In this way, employees who have an affinity with Vox are 11 per cent, while employers who show such affinity are 17.6 per cent and self-employed 14.6 per cent. This allows us to observe that the affinity with Vox is a little greater, in proportional terms, for entrepreneurs and self-employed workers than among employees.

In relation to income levels and the occupation sector (public or private), the data from the CIS Barometer do not show significant differences between the individuals related to Vox and the non-related ones. Both groups of citizens have similar percentages for these variables, confirmed by the chi-square test; therefore, we do not consider them significant for our analysis. However, there are specific traits among those who have Vox affinity according to their level of studies and their socioeconomic status. In connection to studies, we see that the percentage of people with primary education who feel related to Vox is 10 per cent, while those with average studies are 13.4 per cent, and 9 per cent of those with higher education. Therefore, we observe a slightly higher percentage of people oriented to Vox's discourse among those who have medium education, in relation to people with other levels of education. These differences are statistically significant.[8]

Regarding socioeconomic status, we see that almost 9 per cent of the upper-middle class has an affinity with Vox, while 15 per cent of those people who belong to the new middle classes and the old middle classes have a similar orientation. In turn, 11 per cent of skilled workers are favourable to Vox, as are 7 per cent of non-qualified workers. These differences are statistically significant (see chi-square).

Therefore, we see that Vox's speech has more purchase in the middle classes than in the rest of the status categories.[9]

In this way, the base of support for these discursive positions is heterogeneous, but it has a greater proportional significance among the middle classes, males, middle-aged people, self-employed workers and entrepreneurs.

Final reflections

In this last section we simply offer a few final reflections to make a brief assessment of the presented results, and to leave open future lines for deepening analysis of the trends identified. In the first place, Vox's discourse shares some valuations and ideological orientations with different European reactionary movements (and not only European ones). In particular, for Vox it is crucial to strengthen the central state and recover part of its sovereignty ceded to the European Union. Along with this, the defence of Spanish identity is also in line with the rise of European nationalisms. However, Vox does not postulate a discourse of rejection of neoliberalism and financial globalisation, like many other reactionary movements in its environment, and that defend the control of foreign trade and capital. Neither does it have a constant appeal to the significant "people", nor does its policy respond to a logic of articulation of multiple social demands to build a popular identity from an initial multiplicity (in terms of Laclau, 2005). Unlike other European reactionary movements, Vox makes a defence of the fascist heritage of the Franco regime, and the monarchical constitutional order that emerged from the dictatorial period without major ruptures. This can be seen in its frontal opposition to the "historical memory law" that condemns the Franco regime and gives recognition to its victims. The defence and justification of Francoism to justify itself discursively and politically requires keep alive the image of the "enemy". In this sense, the independence movements (Catalan and Basque, fundamentally) take the place of the internal enemy, and for that reason the attack on the independence movement fulfils a central discursive function in the claim of the conservative and centralised social order.

For these reasons, we see similar and different elements with respect to the new European right-wing movements, called "populist" or neo-fascist.

The post-colonial elements that we find in the political logic of Vox, on the other hand, respond to the fears and insecurities that migratory dynamics, and their effect on culture and economy, arouse in sectors of the Spanish population. The discourse focuses on a set of xenophobic and exclusionary proposals aimed at the foreign population. It postulates a series of population control devices, with exclusive effects, to reinforce the sense of security of Spanish identity.

It seems to us that these aspects of post-coloniality present in Vox's discourse are shared by other European movements and are inscribed in the same ideological and political currents. However, the characteristics of the political logic of Vox present marked differences with other experiences, such as those of France,

Germany, Italy, the Netherlands, and so on. It remains to investigate in more depth if these differences respond to the fact of being part of a movement that has an autonomous political logic specific to Spain, or if the particularities of Vox that we have highlighted at the beginning of these final reflections constitute part of a more general European movement. The thread that could link the national "new fascisms" with each other and, thus, embody a new European ghost, would be the will to alter the social order that emerged from the neoliberal EU (and the political and economic pacts on which it is based), in favour of a new equilibrium favourable to the national ruling classes together with a greater discipline and social control of the population, through the political and cultural resources that each national movement has at its disposal. Therefore, to deepen this, it would be necessary to study if these European movements share the will to break or redefine the social pacts in force until the first decade of the twenty-first century, and which are the mechanisms and fundamental mechanisms to carry out such transformations of a social order that produces the sensation of decomposition between a sector of the European middle and salaried classes, and against which the neo-fascist discourses are erected as an alternative that feeds on fear of the future.

Notes

1 The autonomous communities are administrative entities at a territorial level with some legislative and management competences in the territory. The Spanish political division is organised into Autonomous Communities.
2 The chi-square is statistically significant.
3 The chi-square is statistically significant.
4 Both dimensions of the multiple correspondences have an average Cronbach Alpha of 0.64.
5 We have redefined the ideologies that we saw in the previous analysis that are not associated with the sensitivities close to Vox. In the first place, we have regrouped socialism, communism, feminism, social democracy and ecology as ideologies of the "left". On the other hand, liberalism and progressivism are defined as "social-liberal", and we have kept unchanged "Christian democracy" and "conservative" to visualise more clearly their bond with people related to Vox. That is, as a whole, we have regrouped and simplified most of the categories to analyse the specific relationship that conservative ideologies have with Vox's positions. It also merits further clarification regarding those who consider themselves "nationalists". Due to the characteristics of the central hegemonic culture of Spain, we consider that a phenomenon occurs by which individuals, for the most part, associate the "nationalist" option with "peripheral nationalisms", and the Spanish nationalists do not identify themselves with it mainly because they have naturalised. For this reason, it does not appear with high percentages among Vox voters, and it is not significant in the analysis. However, we do use another variable that is different from the barometer that effectively allows us to measure the weight of Spanish nationalist sentiment about the related people of Vox, and which is explicitly analysed in our study.
6 The lower left quadrant is residual, because Vox affinity (high, medium or low) is not found there, nor are the categories of variables that define the democratic dimension. Therefore, only the categories of the variables that define the dimension of the sovereignty of the Spanish state appear there. In the absence of relations of proximity with other categories, it lacks interpretative significance. It constitutes an empty space in terms of the set of correspondences that we analyse.

7 The chi-square is statistically significant.
8 The chi-square is statistically significant.
9 The chi-square is statistically significant

References

Canovan, M. (2005) *The People*. London: Polity.

Castells, M. and Pradera, A. (2018) *La crisis de Europa*. Alianza: Digital, ePub.

Dassonville, R. and Hooghe, M. (2018) 'Indifference and Alienation: Diverging Dimensions of Electoral Dealignment in Europe', *Acta Politica*, 53(1), pp. 1–23.

Eurostat (2018) 'Estructura demográfica y envejecimiento de la población'. Available at: https://ec.europa.eu/eurostat/statisticsexplained/index.php?title=Population_struc ture_and_ageing/es [Accessed 28 April 2019]

Fumagalli, A. (2010) *Bioeconomía y capitalismo cognitivo. Hacia un nuevo paradigma de acumulación*. Madrid: Traficantes de Sueños.

Galcerán, M. (2016) *La bárbara Europa*. Madrid: Traficantes de Sueños.

Laclau, E. (2005) *La razón populista*. Buenos Aires: Fondo de Cultura Económica.

Lorey, I. (2016) *Estado de inseguridad: Gobernar la precariedad*. Madrid: Traficantes de Sueños.

Marginson, P. (2015) 'Coordinated Bargaining in Europe: From Incremental Corrosion to Frontal Assault?', *European Journal of Industrial Relations*, 21(2), pp. 97–114.

Mezzadra, S. et al. (2008). *Estudios post-coloniales: Ensayos fundamentales*. Madrid: Traficantes de Sueños.

Mudde, C. and Rovira, C. (2017) *Populism: A Very Short Introduction*. Oxford: Oxford University Press

Mueller, J.-W. (2016) *What Is Populism?* Philadelphia: University of Pennsylvania Press.

Mouffe, C. (2018) *Por un populismo de izquierda*. Siglo XXI: Buenos Aires.

Prosser, T. (2013) 'Financialization and the Reform of European Industrial Relations System', *European Journal of Industrial Relations*, 20(4), pp. 351–365.

Thompson, P. (2003) 'Disconnected Capitalism: Or Why Employers Can't Keep Their Side of the Bargain', *Work Employment and Society*, 17(2), pp. 359–378.

Thompson, P. (2013) 'Financialization and the Workplace: Extending and Applying the Disconnected Capitalism Thesis', *Employment and Society*, 27(3), pp. 472–488.

Vidal, M. (2011) 'Reworking Postfordism: Labor Process Versus Employment Relations', *Sociology Compass*, 5(4), pp. 273–286.

Visser, J. (2013). 'Wage Bargaining Institutions – From Crisis to Crisis'. Economic Papers 488. European Commission: Brussels.

Documents and files

Centro de Investigaciones Sociológicas (2019) 'Barómetro enero 2019'. Available at: http://cis.es/cis/opencm/ES/1_encuestas/estudios/ver.jsp?estudio=14442 [Accessed 28 April 2019]

Vox (2019) 'Manifiesto fundacional'. Available at: https://www.voxespana.es/manifies to-fundacional-vox [Accessed 28 April 2019]

Vox (2019) 'Programa electoral'. Available at: https://www.voxespana.es/programa -electoral [Accessed 28 April 2019]

5 From Jorge Eliécer Gaitán to Alvaro Uribe

A brief exploration of populism in Colombia

Brett Troyan

Introduction

Before President Alvaro Uribe's election, most historians associated populism in Colombia with Jorge Eliécer Gaitán, the Liberal politician who in the 1930s and 1940s established a large following based on fiery speeches and a direct appeal to the Colombian masses that he referred to as "*el pueblo*".[1] Implicit in the discussion of populism before 2000 was the notion that populism was a movement that was a progressive one albeit one that had authoritarian tendencies because it often relied on one charismatic political figure. For Latin America more broadly speaking the first person that comes to mind is often Juan Domingo Perón. Juan Domingo Perón for most historians of Latin America was a man who championed the working-class agenda in Argentina, bringing about not only a better standard of living for urban working classes, but transforming the traditional way of doing politics as Daniel James so masterfully showed.[2] Jorge Eliécer Gaitán in the 1930s and 1940s revolutionised Colombian politics in a similar manner bringing what he called *el país real* to the attention of the *país formal*. Of course, Jorge Eliécer Gaitán was not the only politician transforming Colombian politics; Laureano Gómez (the Conservative politician) was also an agent of change even if a reactionary one. Populism in the 1930s and 1940s fundamentally altered Colombian politics by challenging the traditional way of doing politics, which was basically an elite affair, and brought new actors such as Jorge Eliécer Gaitán into the political arena.[3]

The current chapter will discuss two key moments of Colombian history when the expression of populism shaped Colombia's political fate. The first key moment was in the 1940s when Colombia witnessed the power of populism with Jorge Eliécer Gaitán's candidacy to the presidency; the second moment was in 2016 when a slim majority of Colombian voters (via a plebiscite) rejected the peace agreement that the Colombian national government had come to with the Fuerzas Armadas Revolucionarias de Colombia/FARC.[4] Most analysts argued that the rejection of the peace agreement reflected the enduring popularity of Colombia's former president, Alvaro Uribe. His political party, Centro Democrático, campaigned actively against the peace agreement and had delegitimised the idea of a peace process through negotiation for several years prior to the plebiscite (Atehortúa, 2016). The current chapter will argue that the continuity between Colombian populism of the 1940s and that of the twenty-first century, despite their very different political agendas, resides in their political practices that delegitimised the existing political system.

Colombia's political exceptionalism in the twentieth century

Colombia's political exceptionalism throughout the twentieth century lay in the enduring power of the bipartisan system; Conservative and Liberal parties were the only real contenders for political power. Jorge Eliécer Gaitán at the beginning of his political trajectory did try to establish an independent political party, UNIR, which failed. His next political step was to transform the Liberal Party from within by democratising the language and appeal of this political party. It would be a mistake to identify Jorge Eliécer Gaitán as the sole democratising element of the Liberal party. Liberal politicians such as Alfonso López Pumarejo brought about universal suffrage for all Colombian men in 1936; and partisan politics in Colombia were quite effective in encouraging the participation of the people in the political system. However, what was different about Jorge Eliécer Gaitán is that he established a direct connection between himself and *el pueblo* whereas traditional Liberal politics were mediated through local party bosses.

Laclau points out that throughout Latin America liberalism and democracy never fused perfectly, but in the case of Colombia Liberalism and democracy were intertwined, which makes the expression of populism in the twentieth century different in Colombia vis a vis other Latin American countries. Laclau states that populism was manifest with the *Estado Novo* in Brazil, Perón in Argentina, and the MNR in Bolivia, which were all political movements or programs that distanced themselves from traditional politics and Liberalism (Laclau, 2009). This departure from traditional politics was seen in Colombia, but as noted above it was still channelled through the traditional political parties. In addition, Colombian populism of the 1940s did not reject Liberalism, but instead embraced it.

Chantal Mouffe explains that populism was viewed often as a political movement that emerges from a pre-modern context; in other words, once a political system or country developed sufficiently populism would no longer be present (Mouffe, 2002). The presence of populist movements from the right in European countries has upended this assumption (ibid., p. 178). Mouffe posits that this explosion of populism from the right should not be seen as a return to pre-modern identities, but as a failure of the traditional political system to present real democratic alternatives (ibid., p. 179). With this political vacuum, populist political parties on the right have emerged to posit themselves as the voice of the people who can defend them from the "elites". Chantal Mouffe states:

> Because the populist parties on the political right are often the only ones who attempt to mobilise passions and to create collective forms of identification. In contrast to all those who believe that political life can be reduced to individual motivations and that political choices are decided by self-interest, these parties are fully conscious that politics always consists of creating an us versus them,

and that political life is mediated through the creation of collective identities. And if their discourse is attractive, it is because it provides these forms of collective identity around the "people".[5]

(Ibid., p. 180)

In the case of Latin America, the emergence of populist movements in the 1930s and 1940s was often attributed to the lack of modernisation of the Latin American political system; political scientists advanced that once Latin American political institutions were better developed and economic modernisation had eliminated pre-modern forms of affiliation, populism would disappear. Some historians have also argued that the Colombian working class was in essence "co-opted" by the Liberal elites in the 1930s and 1940s (Olaya, 2014, p. 141); Colombia's weak modernisation explained why the Colombian working class failed to understand that its political mobilisation within the Liberal Party contradicted its class origins (ibid.). This position is reminiscent of what Mouffe points out as the tendency to describe or explain populism as due to a lack of development either in economic or political terms. As Cristian Acosta Olaya explains, Ricardo López shows how the political identity of *Gaitanistas* was made in the process of their political mobilisation (ibid., p. 148). In other words, Colombian workers did not realise that their common class affiliation united them, but rather that this class identity was constructed while mobilising against the elites (ibid.). Thus, as Ricardo Lopez advances, to see Colombian workers as betraying their class origins is to fail to understand that this class origin was being created during this moment of populism. The debate around co-optation and agency has not gone away when looking at populism on the right in today's politics.

As Mouffe and Laclau emphasise regardless of the degree of economic modernisation and political institutionalisation, people are guided not only by rational considerations, but also by emotions and passions. These emotions and passions become more powerful when linked to the idea of belonging and as Mouffe states, to the idea of an us versus them (Mouffe, 2002, p. 181); Mouffe advances that European populism on the right has capitalised on people's need for a collective identification whereas European leftist traditional political parties have failed to provide this sense of belonging.

Mouffe explains how the drawing of the lines between us and them is often done through morality; she shows how in the case of Austria the populist right was able to capitalise on this notion of the good "native" citizen as the hard-working and moral one versus the so-called lazy and unproductive immigrant (ibid., p. 187). There are also some other interesting parallels between Austria and Colombia with both Austria and Colombia relying on a division of spoils (in terms of political posts and large institutions) for members of the elites of political parties and an essentially bipartisan system. In the aftermath of the Second World War and in the context of the Cold War, Austria's communist party was not allowed to operate freely and was thus pushed out of the political arena (ibid., p. 185).

The two parties that remained significant political players were the Christian Democrat Party and the People's Party (Socialist) (ibid.).

The interesting parallel here is that after the first period of *La Violencia* (1948–1958) a bipartisan system of alternating two political parties (Conservative and Liberal) was put in place from 1958 to 1974. Every four years the presidential post and all political posts alternated between the Conservative and Liberal parties. This political system was known as the *Frente Nacional*; the consequence of this political system was that political passions and real alternatives were in theory not allowed to be presented to the Colombian people.[6] The political game essentially became a client-based one. To understand the shift and the reason for the political bargain one must consider the period of *La Violencia* when passions ran high and, as many historians have argued, there was a breakdown in the civility of the partisan discourse. This breakdown in the political discourse coupled with deeply rooted social inequality and a history of land conflict led to horrific violence, which resulted in the deaths of approximately 300,000 people. In the context of this violence one can understand the acceptance of the political bargain that signified that the democratic electoral process was greatly diminished. The suppression of discourses and processes that challenged this bi-partisan system led in part to the emergence of guerrilla movements.

Populism of the 1940s and Jorge Eliécer Gaitán

Populism of the 1930s and 1940s contributed greatly to the breakdown in civility that eventually led to the period of *La Violencia*, and yet paradoxically also allowed for greater political participation. Herbert Braun has argued that prior to the 1930s, national politics were essentially an elitist affair with major political decisions being made at Bogota's elite Jockey Club by members of the elite of both parties. Jorge Eliécer Gaitán was the most visible figure of Liberal radicalism, but to attribute the popular appeal of radical Liberalism exclusively to Jorge Eliécer Gaitán would be a mistake.[7] Liberalism in Colombia had a long history of appealing to the "masses" that pre-dated the twentieth century. However, Jorge Eliécer Gaitán was the politician who utilised emotional appeals most effectively and who democratised Colombian politics by broadening the stage of where national politics happened and who was included in the political process.

Jorge Eliécer Gaitán was from a background that (as Braun masterfully describes) was neither bourgeois nor working class (Braun, 1985, p. 55).[8] His father engaged in a number of small business pursuits (such as a bookstore) that were unsuccessful and his mother was a schoolteacher (ibid.). His parents' limited means and his complexion that was darker than most elite *Bogotanos* placed him squarely out of the bourgeoisie, but his parents' education and the emphasis on books in his household barred him from working-class affiliation (ibid.). One can speculate that his liminal position in terms of social class and the psychological pressures that resulted from this lack of belonging led in part to his emotionality and to his desire to be noticed. As

Braun reveals, Gaitán was expelled many times from schools and at the school (Colegio de Araujo) that finally accepted him (thanks to his father's Liberal connections) he was known for interrupting classes with his fiery political speeches (ibid., p.56). He went on to become a renowned political orator whose objective was always to bring about continuous participation of *el pueblo* (ibid, p. 57).

Gaitán's choice of profession aligns with Chantal Mouffe's diagnosis of populism; he became a lawyer who defended the unjustly accused. In reality, his cases were not so much about defending the innocent unjustly accused but, as Braun points out, to show the social context that inspired and/or motivated the "crime" (ibid., p. 60). So he saw his role as a moral one in the sense that he was uncovering the true bad actors of Colombia; while the defendants might be on the bench answering for crimes the true perpetrators of injustice were getting off scot free. The other element that is interesting is that law is ideally suited for politics that thrived on this notion of conflict. Mouffe stresses that populism seizes on this feeling of conflict and thus creates a sense of belonging in the expressed antagonism. Gaitán practised criminal law, which is particularly adversarial.

It is worth pausing here to emphasise that the arrival of Gaitán on the scene of Colombian politics did not mean that suddenly Colombian politics became democratic. Colombian politics and the electoral system, although not without its flaws, was fully functioning and democratic in the sense that elections were held. The innovative or singular element here is that Gaitán wanted to encourage participation from *el pueblo* continuously (as Braun points out; ibid., p. 100) and also in spaces other than at electoral polling places. A key ingredient to populism is that the people express their intentions and demands outside of the ballot box and not just through the voting process.

Olaya and Magrini have posited recently in a recent article on *Gaitanismo* that neither populism nor *Gaitanismo* unleashed *La Violencia*, but instead that this political movement acted as a brake on the outbreak of violence (Olaya and Magrini, 2017, p. 296). Olaya and Magrini note that Gaitán took politics to the central plaza/square and rendered visible the masses while performing politics face to face. Much like Daniel James showed how Juan Domingo Perón disrupted the established rules of politics in Argentina by claiming downtown Buenos Aires for the people so too did Gaitán with his massive marches that protested Conservative violence in 1947 and 1948 (ibid., p. 298). Olaya and Magrini see the causes for the explosion of violence as emerging prior to the *Gaitanista* movement. These two authors share John Green's conception of *Gaitanismo* as a populist movement despite its adherence to the bipartisan system (ibid., p. 304). In addition, both Magrini and Olaya underscore the non-violent character of Gaitán's marches in 1947 and 1948 to claim that *Gaitanismo* was a conduit for the people's grievances that prevented the expression of violence because ultimately Gaitán's aim was an electoral victory over the Conservatives (ibid., p. 305).

While I certainly agree that *Gaitanismo* was an organic movement of the working and middle classes and that the expression of populism within the bipartisan electoral system was not indicative of any manipulation and co-optation by the Liberal elites, I view *Gaitanismo* (albeit unintentionally) as an accelerant for the fuel of *La Violencia*. One of the key traits of populism as discussed above is the appeal to emotions and to the us versus them identity formation politics. While Gaitán's agenda was an electoral victory, his mobilisation of the masses was one that showed the power of the masses outside of the ballot box. Meetings in town plazas, extensive and emotional speeches, the writing of letters to his correspondents, and the use of the radio were all meant to inspire emotion and devotion in his political followers. Nowhere is this desire to inspire emotion more evident than in Gaitán's campaign organisation of a "week of passion." (Braun, 1985, p. 106). During this week of passion Gaitán made speeches in various locales and major marches with thousands of *Gaitanista* followers took place in September of 1945, notably the march of the *Gaitanista* torchlight (ibid., p. 108). While these marches were non-violent, their size and the act of taking over of the main public spaces in Bogota showed not only the power of a unified group of people, but also that of its leader. While the workers were non-violent and restrained, the possibility of violence was more than evident. It also demonstrated that Gaitán could summon masses of people and presumably control their public acts. In the process of positioning himself as the true representative of Liberal politics, Gaitán also delegitimised the electoral system as it existed when he ran for presidency in 1944 (ibid., p. 93). Indeed, his rhetoric pointed out the corruption and ineptness of politicians discrediting the political electoral system (ibid., p. 94). Again, there is a similarity with today's populist movements of the right or left that frequently question the legitimacy of existing political institutions. Gaitán's political movement also rejected capitalism or business as usual,[9] which is highly reminiscent of today's populist movements that question the economic order of neoliberalism while blaming immigrants for the growing economic insecurity and austerity that "native" workers face in the globalised world.

Braun emphasises, "His goal was to rebuild the central institutions of society and to return virtue to public life" (ibid., p. 94). While this goal sounds laudable, part of the process of rebuilding the "central institutions" was demolishing their power and legitimacy first. My point here is that Gaitán contributed to the destruction of the legitimacy of the existing political institutions with his rhetoric that imputed that they were corrupt while establishing himself as the political saviour, which signified that with his assassination there were no institutions or organisations that *el pueblo* could turn to. Hence, the only possible response to violence (if one had adhered to Gaitán's view of Colombian political life) was to literally destroy the buildings and the downtown of Bogota that represented the political elites and their "corrupt institutions".[10]

Another point that undermines the view that Gaitán's discourse and marches were non-violent in every sense of the word are the emotions that Gaitán inspired in others, which Braun describes as fright (ibid. p. 95). While the discourse that portrayed and attacked Gaitán was deeply racist and offensive as Braun, Olaya and Magrini point out, Gaitán was breaking new ground by using physicality such as lifting his fist in the air to convey his message (ibid., p. 98). Gaitán's followers also engaged in actions that were violent such as breaking the windows of the newspaper *El Tiempo* to punish the newspaper for not covering Gaitán and his political actions (ibid., p. 100).

Gaitán's political movement in the 1940s was a democratising one that sought to give voice to the ordinary Colombian citizen and to his and her legitimate grievances. It justifiably pointed out the corruption and unfairness of a political and economic system governed by the Colombian oligarchy; however, the political practices that Gaitán used undermined the institutions of the existing political system (which although problematic did ensure elections). In addition, the emotionality of this political discourse and the drawing of the us versus them within the nation are strategies that can be employed very effectively by conservative forces. The abdication of reason to emotionality and the replacement of established institutions with the reliance on a strong leader often signify that ultimately the political system becomes less democratic. However, I would suggest that a populist movement on the left is often a good alternative to the status quo for the majority of citizens even though it may prove to be problematic at a later date. Even though *Gaitanista* followers appeared and were swayed by the emotions that Gaitán encouraged and that the experience of group activities brought about, populism can also be a rational choice (in the short term) in the absence of any real political alternatives to the status quo.

Populism of the twenty-first century and the plebiscite

Populism in Colombia of the twenty-first century, although it brings about the participation of *el pueblo*, into politics is vastly different in its intent and political outlook than populism of the 1930s and 1940s. The aspect of populism that endures in what some scholars have called neo-populism is the performative aspect of populism whether it be of the left or right (Jiménez and Patarroyo, 2019, p. 256). Populism as expressed in the recent plebiscite about the peace accords pitted what many have called populism of the right against political movements of the post-communist left. Perilla Daza states that those who voted yes for the peace agreement were largely members of the indigenous and Afro-Colombian communities, leaders and activists of the LGBT movement, peasants, disabled individuals and victims of the conflict (Perilla Daza, 2018, p. 161). The political parties who supported the peace agreement were Partido Social de la Unidad Nacional (President Juan Manuel Santos's party), Alianza Verde, Union Patriotica, el Polo Democrático, el Partido Liberal, and Cambio Radical (ibid., p. 163). Members of the Centro Democrático, ACORE (Asociacion de Oficiales Retirados de las Fuerzas Militares), el CMA (Centro Mundial de Avivamiento), Cedecol (Concejo Evangelico Colombiano), Fevcol (Federacion Colombiana de las Victimas de las Farc) voted against the peace agreement (ibid., p. 160). In addition to the above sectors and representatives of these sectors, there were Colombian people who were not affiliated to any organisation or party who voted either no or yes to the peace agreement (ibid.). The margin in favour of the no to the peace agreement was razor thin; the difference between the yes and no was less than 1 per cent (ibid., p. 154).

The rejection of the peace agreement was seen as a victory for ex-President Alvaro Uribe who exulted in the aftermath of the plebiscite. Recent scholarship has explored the dynamism and effect of emotions around populism on the right, but the importance of emotions was very much present as well on the left in the aftermath of the plebiscite. Many scholars attributed the strength of the no vote to the emotional appeals made on social media networks that swayed many urban voters to reject the plebiscite. However, as Deissy points out, emotions were displayed on both sides in the aftermath of the referendum results; emotions of sadness, disgust and pain were expressed by those who saw their side defeated in the referendum whereas those who saw the ratification of the peace agreement defeated expressed great joy. The outpouring of grief and disillusionment led to the plebiscite being named as plebitusa, which combined the word plebiscite with *tusa* (which refers to the feeling of rejection one feels when spurned by a loved one) (ibid., p. 173). In these analyses of political decisions such as the plebiscite there is a sense that the power of emotions is greater in the twenty-first century and that somehow the social media networks have exacerbated "irrational discourses" and decisions based on emotionality. Overall, in my view, the emotions that belonging to the Colombian left in this political moment inspired in their followers were of sadness and a sense of betrayal. When I travelled to Colombia in November 2017, many friends and colleagues who belonged to leftist parties or who had worked for decades for peace narrated to me the sense of disillusionment and betrayal of working for *el pueblo* who had seemingly rejected their efforts. A sense of belonging rooted in defeat and despair is difficult to sell; it does not take an expert in psychology to know that political movements are more likely to gain adepts if they give the follower an experience of empowerment and a sense of victory.

The overall consensus about the referendum is that those who came out to vote for the plebiscite were not swayed by rational arguments, but by emotional appeals. After the results of the plebiscite came in, analysts explained that it was largely rural areas that had experienced the violence directly that voted for the peace agreement whereas urban areas voted against this peace agreement (ibid., p. 171). Other scholars such as Andrés Rincón Morera have disputed the characterisation of the plebiscite vote reflecting zones who had experienced conflict versus those who had not (Morera, 2016). Morera explains that within the category of geographical areas where the conflict between the FARC and armed forces was taking place there was no clear consensus around the plebiscite vote (ibid., p. 138). Rather, this author suggests that the areas of Colombia who were inhabited by citizens who wanted a political solution (meaning an end to the war and a political inclusion of the

FARC into the political system) were the ones who voted affirmatively for the plebiscite whereas the no vote reflected the sentiment of citizens who were for a military end to the conflict (ibid., p.139). President Alvaro Uribe's party strongly advocated for a military end to the conflict, which was the view shared by the inhabitants of the geographical zones that voted to reject the plebiscite. Rincón states, "In summary, instead of lightly assuming that the explanation for the voting patterns in the plebiscite lies in a linear consequence of experiencing violence, it seems preferable to examine the way in which political preferences have organized themselves around social factors"[11] (ibid., p. 144). This author's rigorous statistical analysis of municipalities and different modalities of violence (homicide, displacement and armed actions) demonstrates that measuring whether violence increased or decreased between the Uribe government period and Santos's presidency is a better predictor of the voting patterns around the plebiscite. His analysis finds that it was not so much the total number of violent actions that determined whether voters approved the plebiscite, but whether it had increased or decreased from the time period of Uribe's government.

While Rincón does not develop the following point, he also advances that the political narrative around violence and peace was what determined voting patterns (ibid., p. 147). This suggests that populism is particularly powerful for better or worse when it helps people to create a narrative around their reality. Psychologists and psychiatrists who work with trauma victims have shown that one of the key elements to recovery is the construction of a narrative after trauma. Trauma often leaves a victim without a clear sense of events and their sequence; making sense or giving meaning to past events are often essential elements for a psychological recovery. It also makes sense that a simple and clear narrative around who the bad and good guys are (which is what populism offers) would be much more psychologically reassuring than a narrative that is nuanced and where shades of grey exist. Thus, populism as expressed by the Centro Democrático and President Alvaro Uribe provided a reassuring narrative after the trauma that many generations of Colombians have endured during the past eighty years.

An article by Maria Fernanda Gonzalez makes a valuable contribution by examining the discourse (tweets and other messages/texts on social media platforms) that encouraged Colombians to vote against the peace agreement (Gonzalez, 2017). One of the very first major arguments against President Juan Manuel Santos and the peace agreement that Alvaro Uribe and his political allies pushed forward in this plebiscite was that President Santos was a traitor and an unreliable politician since he had promised to follow former president Alvaro Uribe's political platform but instead forged his own path (ibid., p. 118). The second claim made was that this peace agreement turned the country over to the guerrilla army of the FARC (Fuerzas Armadas de Colombia); the FARC were also described as being infiltrated by or agents of Castro-Chavismo (ibid., p. 121). The message from the Santos camp was much more nuanced in that it did not affirm that this was a "perfect agreement", but rather a good enough one.

This nuanced message contrasted with the strong language put forward by the opposing camp that this was a terrible agreement that was dangerous. In addition, the no side was much more vocal with a large number of tweets disseminating its message in contrast to Santos's camp (ibid.). The third claim was based on false claims ("fake news") that the peace agreement promised handsome salaries to ex-guerrillas and that taxpayers (pensioners) would pay an additional tax to fund these salaries (ibid., p. 123). Also, the no to peace side suggested that a new gender ideology (this new gender ideology was in fact just stating the Colombian government's commitment to gender equity) was also a part of this peace agreement (ibid.). In other words Conservatives feared that a peace agreement with the FARC would also bring about "radical feminism". The reactionary discourse and practices of populism on the right is certainly not exclusive to the Colombian right and is very much evident in the case of Brazil as well. The other and final most problematic aspect of the peace agreement (for the side of no) was the insertion of ex-guerrillas members into Colombian political life (ibid.). Gloria Fernanda Gonzalez states, "The government's discourse appealed to logical thinking rather than emotions"[12] (ibid.). In addition, the group opposing peace was able to characterise the peace agreement in reductive and simple ways, which were much more effective than the complex and nuanced statements emitted by the Santos camp (ibid, p. 126). Because much of the ideological war around this peace agreement took place on social networks, short and reductive messages worked best particularly with a public that was predisposed to not like the peace agreement.

Thinking about the role of emotions and reason around the plebiscite agreement, it is clear that on both sides emotions played a role in the voting process. Granted the emotionality of the yes to the peace agreement took place to a much greater degree after the defeat of the plebiscite. Emotions per se are not negative and do not always lead to poor decisions in political processes. Citizens all over the world make decisions based on both rationality and emotions. The populist phenomenon that is problematic in my view is the reliance on the leader who channels and directs the people's will. The takeover of public space (whether this political space consists of political institutions, public squares, and roads/bridges) is a powerful mechanism and strategy for disempowered people to have their voice heard. The example of the indigenous movement based in Southwestern Colombia (Cauca) that routinely occupies the *Panamericana* highway to express discontent and to push national political leaders to listen to their grievances works well and does not necessarily give rise to violence. The key to the enduring power of protest organised by the indigenous communities is that the boundaries of the collective are not fixed; and inclusion rather than exclusion characterises the mobilisation of indigenous communities (for the most part). The *minga* an indigenous term that refers to collective work becomes la *minga nacional/ national collective work* where students and people from all walks of life are invited to participate to protest the national government's decisions. However, collective institutions on the left must be present so that this political mobilisation ultimately returns the decision process back to the ballot box. In the case of the indigenous communities the role of councils (elected representatives of communities) and of the *guardia indigena* (civilian indigenous guards) play key roles in ensuring that the marches or sit-ins remain peaceful and articulate the collective demands. The case of the indigenous communities' political mobilisation in Colombia confirms what Boaventura de Sousa de Santos argues is essential for the democratisation of today's democracies; Boaventura de Sousa advocates for the rank and file of political parties to take a central role in the elaboration of their agenda (Santos, cited in Jiménez and Patarroyo, 2017, p. 261).

In conclusion, populism if defined as a series of political practices that encourage the participation of the masses, can be a force for change either positive or negative. The over-reliance on political leadership, whether it be Jorge Eliécer Gaitán or President Alvaro Uribe, is a dangerous one.[13] While Jorge Eliécer Gaitán certainly fought for the "little guy", there is no predicting how he would have behaved once and if he had acquired political power. As is well known, there were aspects of Juan Domingo Perón's political leadership that were problematic once he gained power. In terms of agency versus co-optation, voters can be manipulated into believing false facts and persuaded to believe in a world view that is exclusionary and racist. However, the boundary between reason and emotion is not always clear. Voters who are swayed by a worldview that encourages them to believe that the "other" is responsible for their decline in standard of living or exploitation may do so because they understand all too well (at some level) that to question the existing structures of power and global capitalism would be much harder. Human beings are very complex in that their motivations and emotions cannot be neatly categorised. I bring up this rather common-sense idea because some social scientists seem to express great puzzlement when contradictory political beliefs are held. For instance, Alvaro Uribe's popularity (he left office with a high percentage of Colombians around 70 per cent who supported him and his administration) cannot be solely explained away by his ability to manipulate truth, but must also be understood in the context of the perception that he had restored "law and order" and that the economy had grown under his presidency. In other words, there were rational pay-offs to believing President Uribe's rhetoric. Also, as Juan Manuel Caicedo Atehortúa demonstrates in his analysis of the *Centro Democrático*'s discourse, populism on the right in Colombia provided a sense of collective belonging in opposition to the "subversive" guerrillas (Atehortúa, 2016, p. 20). In terms of comparing Colombia to the rest of Latin America, I would argue that Colombia has become more similar to the rest of Latin America. The bipartisan system that governed Colombian politics for so long is no longer in existence; multiple parties compete for power. Jorge Eliécer Gaitán started the process of eroding the political system based on the popular mobilisation within the two political parties. The violence that exploded and endured after 1948 gave birth to a political system that was formally democratic, but that excluded real political alternatives. Much of the insurgency that emerged in Colombia from 1964 on can be explained by the closing down of democratic pathways that existed in the traditional bipartisan political system.

While new political alternatives exist in Colombia since 1991, it is not clear how effective these new political movements are in addressing the profound economic and social inequality that permeates Colombian society. Ironically, the breakdown of the traditional political structures while empowering political movements based on different progressive grassroots agendas also gave rise to a reactionary populism. The national *minga* that is taking place in Colombia as of late March and beginning of April 2019 (as I write these words) gives us hope that grassroots movements that come collectively together and that join forces with leftist political parties can bring about positive political change.

Notes

1 It is not my intention to give an overview of the historiography of populism in Colombia. For an excellent and comprehensive analysis of the literature see Olaya (2014).
2 See Green (1996).
3 See Braun (1995).
4 There were other significant moments of populism in twentieth-century Colombia that this essay will not explore. Notably, the political party ANAPO (founded in 1961) led by General Rojas Pinilla (who governed from 1953 to 1957) was a populist one. There was also the MRL (Movimiento Revolucionario de Colombia) that operated within and outside of the Colombian Liberal Party in the 1960s. Both these political movements contested the political arrangement created by the National Front governments.
5 The original quote is in French. This is my translation. Car les partis populistes de droite sont souvent les seuls qui tentent de mobiliser les passions et de créer des formes d'identification collective. Contrairement à tous ceux qui croient que la politique peut se réduire à des motivations individuelles et qu'elle est impulsée par le seul intérêt personnel, ces partis sont pleinement conscients que la politique consiste toujours en la création d'un *nous* opposé à un *eux*, et qu'elle passe par la constitution d'identités collectives. Et si leur discours est si attractif, c'est parce qu'il fournit ces formes d'identification collective autour du "peuple".
6 Most Colombian historians of the National Front period have characterized this political period as one of restricted democracy.
7 See Stoller (1995) for a discussion of radical Liberalism and Alfonso López Pumarejo's contribution to the radicalization of Liberalism.
8 The page numbers that I give for Braun come from the electronic version of his book, which do not correspond to the printed version of the book.
9 See Braun (1995, p. 93) for Gaitán's take on capitalism.
10 This is what happened in the *Bogotazo*.
11 "En suma, antes que imputar ligeramente la explicación de la votación por el 'Sí' y por el 'No' en el plebiscito como consecuencia lineal de la violencia, pareciera preferible contemplar la manera en que las preferencias políticas se han venido organizando en relación con factores de orden social.' This is the original quotation in Spanish. It is my translation.
12 "Los mensajes del gobierno apelaban más a la razón que a las emociones." Original quote in Spanish. This is my translation.
13 De Sousa has emphasised how ultimately popular political participation is delegated to a leader. Cited in Jiménez and Patarroyo (2019, p. 261).

References

Atehortúa, J.M.C. (2016). 'Esta es la paz de Santos? el partido Centro Democrático y su construcción de significados alrededor de las negociaciones de paz', *Revista CS*, 19, pp. 15–37. DOI: http://dx.doi.org/10.18046/recs.i19.2136

Braun, H. (1985) *Assassination of Gaitán: Public Life and Urban Violence in Colombia*. Madison: University of Wisconsin Press.

Braun, H. (1995) *The Assassination of Gaitán: Public Life and Urban Violence in Colombia*. Madison: University of Wisconsin Press.

Gonzalez, G.F. (2017) 'La postverdad en el plebiscito por la paz en Colombia', *Nueva Sociedad*, 269, pp. 114–126.

Green, J. (1996) '"Vibrations of the Collective": The Popular Ideology of Gaitánismo on Colombia's Atlantic Coast, 1944–1948', *Hispanic American Historical Review*, 76(2), pp. 283–311.

Jiménez, J.A. and Patarroyo, S. (2019) 'Populism in Democratic Contexts in Latin America: Review of the Empty Signifiers in the Discourse of Three Populist Leaders. A Study from the Political Analysis of Discourse', *Revista Mexicana de Ciencias Políticas y Sociales*, pp. 255–288.

Laclau, E. (2009) 'Laclau in debate: Post-Marxism, Populism, Multitude and Event' (Interviewed by Ricardo Camargo), *Revista de Ciencia Política*, 29(3), pp. 815–828.

Morera, A.R. (2018) 'De la esperanza a nuevas incertidumbres. Sobre la distribución de la votación en el plebiscito colombiano (2016)', *Análisis Político*, 31(92), pp. 137–158.

Mouffe, C. (2002) 'La fin du politique et le défi du populisme de droite', *La Découverte/Revue du MAUSS*, 20, pp. 178–194.

Olaya, A.C. (2014) 'Gaitánismo y populismo: Algunos antecedentes historiográficos y posibles contribuciones desde la teoría de la hegemonía [Gaitánism and Populism: Historiographic Backgrounds and Possible Contributions from the Theory of Hegemony], *Colombia Internacional*, 82, pp. 129–155. DOI: dx.doi.org/10.7440/colombiaint82.2014.06

Olaya, A.C. and Magrini, A.L. (2017) '"Cursed Words", Gaitánism, Violence, and Populism in Colombia', *Papel Político*, 22(2), pp. 279–310.

Perilla Daza, D.C. (2018) 'La plebitusa: movilización política de las emociones posplebiscito por la paz en Colombia', *Maguare*, 32(2), pp. 153–181.

Stoller, R. (1995) 'Alfonso López Pumarejo and Liberal Radicalism in 1930s', *Journal of Latin American Studies*, 27(2), pp. 367–397.

6 The social question in the twenty-first century

A critique of the coloniality of social policies

Angélica De Sena

The social question and the politics of sensibilities

The entire twentieth century can be reconstructed through the efforts of the state to diminish the disruptive potential of conflicts, of the various regimes of accumulation, through social policies. Basically, the state has intervened in the capital/labour conflict and its consequences with the creation of systematic bureaucratic mechanisms that provided workers with goods and services aimed at stabilising potential disruptions. From the Beveridge Report, through the ideas of Titmuss and the practices of the state of Keynesian welfare to the "massive programmes" of today, the United States and Europe have created mechanisms of compensation for the consequences of the conflicts generated by capitalism and its expansion. On more than a few occasions, these "styles" of social policy have been called populist. One of the central characteristics of social policies is that they are instruments through which sociabilities are elaborated, sensibilities are built and the basic nodes of the political economy of morality are reinforced.

In this context it is possible to understand how one of the strategies most used in the twentieth century to manage and weld the cracks left by the tensions and conflicts between capital and work has been the so-called welfare state. In this strategy, the centrality of social policies has been fundamental according to spatial and temporal contexts. Beyond the different conceptualisations of each modality of the welfare state, social policies have had a way of managing the images of the world and sensibilities of those that can only be "assisted" so that they do not constitute a conflict.

In this regard, we mention the example of the United States, where the New Deal during the presidency of Franklin Roosevelt (1933–1945) can be located as the beginnings of the welfare state with the aim of supporting the poorest sectors of the population. The aim was to reform the markets and revitalise the American economy strongly hit by the Great Depression of the 1930s. The measures taken aimed to generate economic recovery, and for this purpose public works were undertaken to create employment opportunities, and in 1935 the Social Security Law was passed establishing a social protection system at federal level with retirement for over-65s, insurance against unemployment and various aids for the handicapped. Progressively, the system came to cover an ever greater part of the population, particularly thanks to the amendments of 1939 and 1950.

In Europe, in comparison, in 1942 the first concrete form of the welfare state was produced with the Beveridge Report, which identified the five great evils of British society: squalor, ignorance, want, idleness and disease; and it was essential to act upon these. In this context, the state should ensure interventions among the population towards the construction of social protection designed for the entire life of the citizen, attending to the economic risks.

In Argentina, towards the end of the nineteenth century, social hygienism, of positivist origin, gave way to medicine in welfare policy as the scientific way of containing social problems. In reference to charitable societies, Moreno claims that these institutions were "contributing to justify 'scientifically' the control over the individuals interned in the establishments under their responsibility" (Moreno, 2013, p. 8). This allowed the state control over what was considered a dangerous crowd of hungry people.

As for the public sphere in the city of Buenos Aires in 1890, it worked through municipal jurisdiction, offering some assistance to the "poor of solemnity" linked essentially to medical assistance and the prophylaxis against infectious-contagious diseases.

Over the years, and throughout the history of labour and industry, the world was modelling the social issue. In Argentina the charitable societies were diluted as such, with the state taking primary control over social issues, but also at the end of the twentieth century so-called civil society organisations or NGOs grew in the world order, with a high participation in the design and implementation of social policies. This leaves the question regarding the issue of poverty as a social problem that admits the possibility of state assistance or private welfare actions (Grassi, 2000). In this way it is possible to use a definition of social policies as a combination of public and private; then while the state is engaged in "combating" poverty, the individual finds himself alone and, in some sense, "responsible" for his situation.

Public policies can be characterised as a form of organisation and administration of a state that refers to courses of actions and/or omissions aimed at an end, as a result of a decision process in relation to bids of interests and conflicts between agents (individuals), agencies (institutions) and discourses (interaction between agents and agencies) (Medellín, 2004).

In this way, the welfare state was born as an instrument to sustain the balance of the complex social interactions of a society in which conflicts of a different nature and different agents are constantly woven, which form forces and social tensions that contribute to the definitions of the type and modality of social policies. Esping-Andersen (1993) states that social policies can be "emancipatory", according to whether they break (or not) with the dependence of the subject on the state, as well as "legitimising", as long as they do not contradict or help the market processes, holding *the state* of things within the system. But always the various policies designed and implemented by capitalist states respond to the model of *benefactor, corrector* or *compensator* of social inequalities (Esping-Andersen, 1993).

However, the entire twentieth century can be reconstructed through the efforts of the state to diminish the disruptive potential of conflicts, of the various regimes of accumulation, through social policies

So, from sociology, beyond the different conceptualisations of each modality of social policies, they have had a way of managing the images of the world (Scribano, 2004) and sensitivities of those to whom it is only possible to "attend", while state practices that perform the social with a capacity to build sociability. In this sense, our studies show the preponderance of the adjective of "all" politics as "social", thus endowing it with a certain character of "positive" valuation to the state action for which it is reserved directly/indirectly, and in this way, the ability to compensate for market failures and civil society with respect to inequality (De Sena, 2014). In the same direction, we have pointed out that we can verify the existence of a "hidden curriculum" (metaphorically returning to the concept used in the analysis of teaching practices) of social policies through which devices for the regulation of sensations are constructed, devices that strengthen the looks that carry the images of the world that they suppose (De Sena, 2014). At the same time, we have made enquiries about the relationships between sensitivities and hunger management (Scribano, Huergo and Eynard, 2010; and Eynard, 2011), regarding the states of the politics of sensations (Scribano, 2008), and the characteristics of some societies structured around enjoyment (Scribano, 2013). Along with this, we have explored the consequences of various educational practices in the lives of poor women, such as the "dissolution/transformation" of educational policies in care-oriented policies. The reconfiguration of the cognitive in the specifically affective elements of all school practices indicates that what is "apprehended" is to feel according to gender, age and class position, so that in societies normalised in the act of immediate enjoyment through consumption they seem to establish the ways that "sensitivities will allow them to endure" (Scribano and De Sena, 2014). We have found that social policies, in relation to the group of people whom they aim to "assist", act indefinitely and keep them occupied, thereby constituting *occupability* as a new way to repair market failures and avoid conflict (De Sena, 2016). This is because, the politics of bodies and emotions are inscribed and elaborated in certain geopolitical and geocultural contexts. In the current situation of the Global South we can partially characterise this context by understanding the transformations of two of its most important edges: the social regime of accumulation and the political regime in which they are developed. The first refers to a set of economic, social, cultural and legal institutions through which the process of production, distribution and accumulation (reproduction) of goods and material values of a society takes place. The political regime can be understood, as the set of institutions and processes, governmental and non-governmental, carried out by social actors endowed with a certain capacity for power, through which the political domination of society is constituted and exercised. In generic terms, "politics" as state praxis refers to a set of actions developed in a planned manner and following deliberately designed strategies in pursuit of the objectives that are intended as results of the same. And it is the state, which – because of its coercive capacity and the general scope of its intervention – is constituted as the political institution par excellence and the main executor of policies aimed at producing effects that involve society as a whole, contributing in such a way to decisively model the public (and private) scope of it. Consequently, when we speak of "policies" in a more limited sense, reference is made more specifically to a normative and/or executive attribution of a general and public order of competence, in principle, exclusive of the state. This means that their analysis inevitably refers to the accumulation model in force in each society, and this makes it possible to understand that the policies implemented decades ago are favourable formulas for the regulation of social aspirations and conflicts (Halperin Weisburd et al., 2008). But also, it is necessary to observe that this regulation advances on the bodies, emotions and actions of each one of the people that make up the social groups in each city, this shapes and consolidates ways of life, of doing and perceiving, that organise the feeling of the populations.

Inscribed in this context it is possible to understand (at least partially) that social policies fulfil a main function: that of attenuating conflicts that occur in different classes or social groups, this being the terrain where, in a privileged way, they connect with the politics of bodies and emotions (De Sena and Cena, 2014; Scribano, 2012). In a double sense, the concrete ways of distribution (circulation and accumulation) of social plans imply a set of practices associated with the policies of the bodies and also a set of ideological practices associated with the politics of the emotions tending to diminish and/or delete something.

Any critical reflection on the social question implies much more than the immediate problematisation of poverty – or what at that particular moment is being expressed as a social problem; it also involves the thematisation – and significance – of related problems such as the explanation of the situation of unemployment, job insecurity, and so on. From this perspective, all social policy (and public policy in general) is traversed by a particular politics of emotions that will shape the ways in which actors in conditions of denial feel, experience and act in contexts of poverty. If, as I mentioned earlier, social policies occupy a central place in guaranteeing the reproduction of the regime, the policies of emotions allow us to begin to elucidate some of the strategies – presented as the most intimate, individual and subjective – of the regime for its reproduction at the expense of an increasing number of the population living in conditions of denial and that do not represent a threat to systemic ends.

Perceptions, sensations and emotions constitute a tripod that allows us to understand where sensitivities are based. Social agents know the world through their bodies. What we know about the world we know by and through our bodies. In this way a set of impressions impact on the forms of "exchange" with the socio-environmental context. The impressions of objects, phenomena, processes and other agents structure the perceptions that subjects accumulate and reproduce. Constituted thus is a naturalised way of organising the set of impressions that are given in an agent.

The sensations, as a result and as antecedent of the perceptions, give rise to the emotions, which can be seen as the puzzle that comes as an action and an effect of feeling or feelings. They are rooted in the states of feeling the world that allow us to sustain perceptions associated with socially constructed forms of sensations. In turn, the organic and social senses also allow the establishment of a vehicle that seems unique and unrepeatable as are the individual sensations, and to elaborate the "unnoticed work" of the incorporation of the social made emotion.

Thus, the politics of bodies, that is the strategies that a society accepts to respond to the social availability of individuals is a chapter, and not the least, of the structuring of power. These strategies are knotted and "strengthened" by the politics of emotions tending to regulate the construction of social sensitivities.

The politics of emotions require regulating and rendering bearable the conditions under which order is produced and reproduced. In this context, we will understand that the mechanisms of social support are structured around a set of practices made body that are oriented to the systematic avoidance of social conflict. The devices for regulating sensations consist of processes of selection, classification and elaboration of socially determined and distributed perceptions. Regulation involves the tension between senses, perception and feelings that organise the special ways of "appreciating-in-the-world" that classes and subjects possess.

The social support mechanisms of the system do not act either directly or explicitly as "control intent" or "deeply" as focal and punctual persuasion processes. They operate "almost unnoticed" in the porosity of custom, in the frameworks of common sense, in the constructions of sensations that seem the most "intimate" and "unique" that every individual possesses as a social agent.

Among them there are two that, from a sociological point of view, acquire relevance: fantasies and social ghosts. Some are the reverse of the others, both refer to the systematic denial of social conflicts. While the fantasies occlude the conflict, they invert (and consecrate) the place of the particular as a universal and make the inclusion of the subject impossible in the fantasised lands, the ghosts repeat the conflictual loss, remember the weight of the defeat, devalue the possibility of the counter-action in the face of loss and failure. Fantasies and ghosts never close, they are contingent but they always operate, they become practical. Thus, they constitute "practices of feeling" that update/embody in concrete processes the set of sensitivities that constitute the politics of emotions.

Public policies when creating sociabilities also build experientialities and sensitivities in such a way that what is shared unnoticed by management practices is embodied. The social-made body is knotted and woven with the incorporated statehood, thus including in the life of the subjects a certain experience coming from the results of the dialectic between state practice and social practices.

In close connection with the above, and as a metonymic expression of the phenomenon, there is a strong link: the practices of statehood are related to the practices of a normalised society in the immediate enjoyment through consumption. The explicit intention of the economic policies of the current democracies is to seek growth by increasing domestic consumption where the mass of it plays a role of fundamental importance.

The strong connection between economic policy, social policy and the market becomes a restructuring factor of sociabilities and creates the conditions of possibility so that the immediate enjoyment in and through consumption becomes experience. The politics of emotions, insofar as they harbour practices of feeling, are crossed/permeated by the consequences of the dialectic that is updated between images of the world included in public policies and the sensitivities constructed by the alluded policies of emotions. The contradictions that exist between citizens, consumers and bearers of rights are embodied in millions of people and, in this instance, differential forms of sociability, experientiality (*sensu* Scribano) and sensitivity are crossed.

Social conditional cash transfers as a mode of global intervention

In this section I will try to show the ways in which the capitalist states today have decided to intervene in the so-called "vulnerable" population, beyond the type of country and forms of government, based on the increase in conditions of poverty and unemployment in the global order. The states take sides by the implementation of a particular type of social programmes: conditioner cash transfer (CCT).

On the one hand, the twenty-first century incorporates the idea of inclusive social policies (Arroyo, 2006), with a favourable view of massive programmes for poverty and incorporating the concept of *universalism*, occluding that it is not for everyone, evidencing that it leaves aside part of the total excluded population, enabling the persistence of guiding criteria for the "eligibility" of the subject (De Sena, 2011). It is in this sense that the advance of the multiple forms of the devices to intervene on the welfare of society allowed the appearance of the so-called conditioner cash transfer as an intermediate modality between cash and service transfers, given that in some cases this refers to money in cash, in others to pre-filled cards for the purchase of certain products or directly from goods (in general food) or directly in kind. It is possible to locate these types of devices "imported" from the United States since their origin is located there, in different forms such as transfers for the purchase of food in the 1930s, and the choice of schooling for children in the 1970s, among other forms.

The CCT programmes (promoted in recent decades by the main multilateral credit agencies), beyond journalistic or propaganda information, do not mean a profound transformation in situations of poverty; they are, for a long time now, one of the mechanisms selected by the capitalist state to facilitate, improve and guarantee its reproduction over time of the poorest sectors of society (De Sena, 2016).

These types of programmes, implemented in the global order, from the perspective of Cecchini and Atuesta (2017) on the one hand refer to the co-responsibility of the recipient subject, and on the other hand they are part of the innovation of social policies aimed at overcoming poverty and "they have managed to cover populations traditionally excluded from any provision of social protection, articulating different inter-sectoral actions – especially in the field of education, health and nutrition – from a multidimensional perspective. The CCT have also been characterised by their 'innovative management model'" (Cecchini and Atuesta, 2017, p. 9). In Latin America, they have been implemented for more than two decades and "have meant a break in relation to the traditional clientelistic mechanisms of social policy [...] have contributed to modernizing social policy through technological innovations, such as the introduction of records of recipients and computerized management systems [...] under certain conditions that seek to improve the human capacities (mainly in education and health) of its members, especially children and adolescents". In this way, the CCT seek reduce poverty in the short term through direct monetary transfers, which allow basic levels of consumption to be sustained, and in the long term through improvements in the levels of health and education of children from poor households. Several impact evaluations have shown that CCT have managed to improve the welfare of the poor population in various aspects such as income, food consumption and access to education and health, among others (Cecchini and Atuesta, 2017, pp. 9–10).

Meanwhile, a study carried out by the Inter-American Development Bank (IDB) affirms that, although these programmes have "revolutionized social assistance throughout the world", the beneficiaries of these programmes continue to be poor, with low education and job instability, of course. That possibly the situation of these people would be even worse without these programmes (Stampini and Tornarolli, 2012).

These social programmes are not limited to Latin America, they can be tracked in the US through forms of coupons to "help" families with some type of lack, related to food, housing and education, among others. This is also the case in Europe with the case of France with the Revenu minimum d'insertion (RMI), or in Italy with the Voucher Famiglia, among other countries and programmes. In all cases, they give an account of the needs of the family or individual that requires some kind of help at some point in their life stage, which may refer to "social inclusion"; but there must always be a consideration on the part of the beneficiary subject (De Sena, 2016).

In these contexts, in the last decades on a global scale there have been massive interventions by the state. Massive in the sense of more social programmes and more people receiving them. These interventions are social cash transfer as a way to achieve inclusive development; that is, giving money the power to reduce poverty and inequality, empower women, prevent conflicts, consolidate peace, access health services and education and foster "social inclusion" (DFID, 2011; Soares, 2012; Barrientos et al., 2013). Currently, CCT have spread worldwide with a strong presence in Latin America, Asia, Africa, Turkey, Europe, India, the United States, and this number and variety of countries allows us to ask ourselves: What is the meaning of these interventions? What are the results? Next, I will analyse some experiences on a global scale of these interventions by the state, in order to hypothesise some answers.

In the case of India, the Public Distribution System is an in-kind food programme based on the transfer in kind with an objective of equity, which provides food grains cheaply to the poor, that works on a massive scale. On the one hand it has undergone radical changes since its inception in 1939, linked to the changes in government policy, and with each political change there have been multiple implementation problems with their consequent impacts on welfare and changes in the demand for subsidised grains, in a context of concern for food security. This programme is demanded by working households in rural areas that do not own land, with greater vulnerability in the times that govern forms of food availability and inflation affects their price. Historical evidence marks the negative effects of systemic changes. On the one hand, the programme has been criticised by various academic and planning agencies for being wasteful and redundant and, along with this, there are pressures to convert it into an income transfer scheme, following the experiences of other developing countries; but on the other hand, they consider that it is necessary to expand the same schemes to guarantee the alimentary hygiene of homes. This programme supports producers and consumers and enables access to food for the population, which is why it is important to provide continuity. Throughout the decades the programme underwent modifications, but this does not entail its discontinuation; the discussions are only in the order of how to best continue it (Bhattacharya, 2018).

Noting that very diverse situations are found in other continents, I will take two countries in Africa: Burkina Faso and Ghana.

Burkina Faso was one of the first African countries to adopt this type of programme, called Orphans and Vulnerable Children. With this type of intervention, positive effects on poverty are recognised in the short term, but in the context in which they are implemented, it is unlikely that sustainable changes will occur in the long term. They also require a thorough review of the role of men and targeting, in contexts where poverty is widespread, therefore it is not unlikely to have adverse impacts in terms of social cohesion, inequality and poverty (Malmi, 2018).

In the case of Ghana, we see implementation of various poverty care programmes such as the National Health Insurance Scheme and the Ghana School Feeding Programme, together with the Income Transfers Programme Livelihood Empowerment Against Poverty, that have allowed some reduction in poverty; beneficiary households begin to participate in productive activities such as agriculture and animal husbandry; there is improved access to food, education and health; and above all they contribute to peaceful coexistence and social cohesion, although challenges remain. Although cash transfers' virtues are recognised, they also show situations of certain conflict between those who are effectively recipients and those who are not (Alatinga, 2018), as a consequence of the programme's focus: "For example, MacAuslan and

Riemenschneider (2011) study on cash transfer in Zimbabwe and Malawi reported on resented, conflict and jealousy arising from the divisiveness of some people being targeted while others are not, even though they are all considered poor" (Alatinga, 2018, p. 56).

In Europe the fight against poverty follows the same rules. Such is the case with Spain – which is considered "a poor country among the rich and rich among the poor" – and to address this situation the answer has been the National Plan of Social Inclusion, the most recent iteration of which refers to the years 2013–16 and which is intended to respond to the needs arising from poverty and social exclusion. This programme consists of 240 proposals for measures that are articulated around the strategic objectives of the active inclusion approach promoted by the European Commission. Among these proposals is "assistance through minimum income policies that ensure coverage of basic needs". This income system is developed through the Autonomous Communities through benefits in situations of need such as social emergency, child poverty, non-contributory and welfare pensions, subsidies and aid of an assistance nature for unemployment, and family benefits. At present all the Autonomous Communities have some programme of this type, although there is no proper system of income guarantee in situations of lack or insufficient resources (Pastor Seller et al., 2018).

The Italian experience is not too far away, although Italy and Greece were for many years the only countries in the European Union among the 27 that did not have a national measure of monetary income. Then, in a process initiated in 2012, there was brought into effect in March 2017 the approval of a law to establish the Income of Inclusion in Italy. In this way it is possible to find CCT (for example subsidies) and offer of assistance services (transfers for education, health care, etc.) activities and the so-called *voucher*, considered for the economic sciences as an intermediate modality between transfers in cash and services, as a way to improve welfare. The voucher is located as an instrument of access to services in the field of a *welfare* that seeks to imitate market structures in order to make less expensive interventions disseminated in a capillary and qualitatively more satisfactory way of guaranteeing compensatory-type benefits under the guise of universality by postulating subsidies as an operational principle of social policies. These are implemented in relation to the educational issue, the fight against poverty, health care for the elderly and disabled, the care of babies and children, training for work and job placement, being crystallised as "contributions" possible to standardise as forms of "help" (De Sena, 2016).

Moving to another continent: in Latin America, the ways of intervening in the face of poverty are not so different. Once again they are oriented around conditional cash transfers, and under that nomination they are located and entail focused and progressively massive money transfers to the populations. The beginnings of the CCT go back to the beginning of the 1990s when some countries had implemented them: in Mexico, initially called Progresa in 1997, their budget for 2002 reached US$1.9 billion, covering almost three million rural families and more than 1.2 million urban and semi-urban families, as indicated by FAO on its website; and today Oportunidades, which were followed by Honduras with the Programa de Asignación Familiar and Nicaragua with the Red de Protección Social, thus being the first Central American countries to start a programme of monetary transfers in 2000. These experiences were followed in Latin America with the Familias en Acción programme in Colombia; Programme for Advancement through Health and Education in Jamaica, Avancemos in Costa Rica; Chile Solidario in 2002; in Brazil with the Borsa Familia in 2003 (in 2017 with almost 12 million families according to World Bank data); the Development Bond Humanin Ecuador, Asignación Universal por Hijo (2009) and Ciudadanía Porteña (2005) in Argentina; the Programa de Apoyo a Comunidades Solidarias (former Rural Solidarity Communities or Solidarity Network) in El Salvador; Abrazo in Paraguay; Juntos in Peru; and Progresando con Solidaridad in the Dominican Republic. Currently, ECLAC notes that they are implemented in 23 countries in Latin America: Argentina, Belize, Bolivia, Brazil, Chile, Colombia, Costa Rica, Cuba, Ecuador, El Salvador, Guatemala, Haiti, Honduras, Jamaica, Mexico, Nicaragua, Panama, Paraguay, Peru, Dominican Republic, Trinidad and Tobago, Uruguay and Venezuela. In 2013, around 137 million people in 17 countries in Latin America and the Caribbean received transfers that represented on average between 20 and 25 per cent of their family income (Ibarraran et al., 2017), making it clear that they are the "star" programmes of the region.

This global journey allows us on the one hand to observe that this type of intervention responds to the way in which historically the different branches of protection have been created, around the notion of "risk" (typical of the insurance contract) that traditionally supported the gradual emergence of the different social protection branches of a contributory nature, but in recent decades has increased the benefits of a character not adding the need for concurrence of another risk: poverty (Pastor Seller et al., 2018). And, on the other hand, it shows the current global structure of capitalism where a metamorphosis of the mode of intervention by the state takes place. On the one hand, the intervention changes, leaving aside the indirect ones (such as via education or health), and the constant increase in the population that receives money and therefore under subsidy from the state is consolidated. That is why these programmes redefine a particular way of linking the receiving subjects and the state, giving shape to a type of social bond. In this sense, the journey made it possible to mention that these CCT range from the transfer of grain as a means of feeding, and which only reproduce the life of those assisted, to cash interventions ensuring consumption by building "new" and more consumers. In both cases we are far from the consideration of a citizen and an autonomous subject. In all these programmes, the analysis, both those of a friendly tone and the most critical ones, account for the fact that these interventions, in one way or another, are a help but not enough; for this reason, this type of device mitigates situations of poverty and places the subject in an assisted situation.

In this regard, society in the twenty-first century must reflect on whether this continues to be a way of liquefying conflict situations or simply the new social question: millions of consumers, poor and helped people.

Conclusion: social policies, populisms and coloniality

As was stated at the beginning of this chapter, it has been possible to visualise, at least partially, the ways in which a planetary process of colonisation of sensitivity can be observed. It has been exposed how compensation policies associated with populisms are today part of the basic forms of state intervention to manage tensions and reduce conflicts. And these same policies become oriented to the management of bodies and emotions in and through sensitivities. With this argument it is possible to identify the global displacement from those interventions of a punctual and temporary nature to the permanent application of palliatives.

In the context of the above, some basic axes can be reconstructed that allow the connection of social policies, populism and coloniality in the current situation of the expansion of capitalism.

In the first place, it is obvious that the tendency to mass social policies as a feature of current populisms responds to the need to guarantee social peace and protect national states from the precariousness of globalisation. The instability of extractive processes and the volatility of financial capital are balanced and "softened" by the massive application at the local level of social plans for conflict management.

Second, it is easy to detect the consolidation of the main effects of social policies in relation to the politics of international sensitivities: State assistance serves to mitigate (hunger, disease, etc.). In this way the medicalisation of state action that moves between the analogy of containing the disease and/or mitigating its effects is reproduced. Keeping the vast majority of people in a situation of poverty happy involves isolating the possible effect of discontent and expansion, as what is not contained can spread. The affliction is mitigated but the causes of it are not addressed, in the way in which a sensitivity policy is consolidated.

Third, it is evident how the extension of social policies conforms to their global application, and that they generate the "replacement" of the centrality of the intra-class distributive bids proper of the welfarist strategies for the intra-class conflicts by the appropriation of social plans. The capital–labour structural conflict for the appropriation of increasingly important portions of resources is complemented and broadened by disputes between the assisted, by conflicts among the poor, and by inter-class antagonisms for the management and accumulation of benefits and social plans.

Fourth, it is possible to confirm the existence of a geopolitics of social plans that, in their planetary distribution, differentiate themselves by the location of those oriented to encourage consumption and those to alleviate hunger. Each country and region has "its" dose of contention according to what is lived in the South or the global North thus arming a continuum with two poles of reference: one constituted by the spaces most linked to the transformations and dynamics of the system (be like losers or winners) towards which the policies of consumer promotion are oriented as a necessary link between economy and society, and the spaces disconnected from the logic of the state-dependent market through which practices are oriented towards hunger.

Fifth, state interventions promote and reproduce the conditions of acceptability of the given: "for something they serve", "they reach for something". Between resignation and indifference, the plans only cover the minimum, they only aim to be a minimum floor; that is, they do not solve the problems to which they refer. The plans in everyday life play a fundamental role: in many places without them there would be no help. The beneficiaries repeat almost a litany of the neo-colonial religion (Scribano, 2013): of *course it serves, it is not enough, but it is an aid.*

In this context, the politics of sensitivities can be drawn as a pentagon of sensations and emotions: a pacified individual, softened, mitigated, contained and relieved. A geometry of the effects of policies that make collective action predictable and conflicts manageable.

If the multiple forms of populism promote state intervention in connection with the problems of sectors not considered (not "included" in the context of the market, etc.), such action produces and reproduces the global expansion of assistance that contains individuals within the walls of dependency, and with it inaugurates a deep phase of coloniality via the politics of the sensitivities.

Bibliography

Alatinga, K. (2018) 'Las Transferencias Sociales de Ingreso para el Desarrollo Inclusivo: un análisis de los actores involucrados sobre las virtudes y desafíos del Programa Livelihood Empowerment Against Poverty de Ghana', in A. De Sena (ed.) *La intervención social en el inicio del siglo XXI: transferencias condicionadas en el orden global.* Buenos Aires: ESEditora, pp. 49–76. Available at: http://estudiosocio logicos.org/portal/la-intervencion-social-en-el-inicio-del-siglo-xxi-transferencias-con dicionadas-en-el-orden-global/. [Accessed 3 March 2019]

Arroyo, D. (2006) La política social ante los nuevos desafíos de las políticas públicas. Centro de Documentación en Políticas Sociales. Gobierno de la Ciudad de Buenos Aires. Documento No. 36. Available at: http://estatico.buenosaires.gov.ar/areas/des_ social/documentos/documentos/36.pdf. [Accessed 20 March 2012]

Barrientos, A., et al. (2013) '"Growing" Social Protection in Developing Countries: Lessons from Brazil and South Africa', *Development Southern Africa*, 30(1), pp. 54–68.

Bhattacharya, R. (2018) 'El "Public Distribution System" en India: Un programa alimentario en especies basado en la transferencia de granos', in A. De Sena (ed.) *La intervención social en el inicio del siglo XXI: transferencias condicionadas en el orden global.* Buenos Aires: ESEditora, pp 19–48. Available at: http://estudiosociologicos. org/portal/la-intervencion-social-en-el-inicio-del-siglo-xxi-transferencias-condiciona das-en-el-orden-global/. [Accessed 20 March 2019]

Castro-Gómez, S. (2000) 'Ciencias sociales, violencia epistémica y el problema de la "invención" del otro', in E. Lander (ed.) *La colonialidad del saber: eurocestrismo y ciencias sociales. Perspectivas latinoamericanas.* Buenos Aires: CLACSO, pp. 145–161.

Cecchini, S. and Atuesta, B. (2017) *'Cash Conditional Transfer Programs in Latin America and the Caribbean'.* Social Policies Series. 224. Santiago: ECLAC.

Demertis, N. (2006) 'Emotions and Populism', in S. Clarke, P. Hogget and S. Thomson (eds) *Emotions, Politics and Society.* New York: Palgrave Macmillan, pp. 103–122.

DFID (2011) Department for International Development Annual Report and Accounts 2010–11 Volume I: Annual Report. Available at: https://assets.publishing. service.gov.uk/government/uploads/system/uploads/ attachment_data / file / 67477 / Annual-report-2011-vol1.pdf [Accessed 2 March 2019]

De Sena, A. (2011) 'Promoción de Microemprendimientos y políticas sociales: ¿Universalidad, Focalización o Masividad?, una discusión no acabada', *Revista Pensamento Plural*, 4(8), Enero–Junio, pp. 36–66.

De Sena, A. (ed.) (2014) *Las políticas hechas cuerpo y lo social devenido emoción: lecturas sociológicas de las políticas sociales.* Buenos Aires: ESEditora/Universitas.

De Sena, A. (2014) '*Notas sobre lo social como ámbito de debates no cerrados*', in P. H. Martins, M. de Araújo Silva, É. L. de Souza Leão and B. F. Lira (eds) *Guía sobre postdesarollismo y nuevos horizontes utópicos.* Buenos Aires: ESEditora. E-book. Available at: http://estudiosociologicos.org/portal/guia-sobre-post-desarrollo-y-nue vos-horizontes-utopicos/. [Accessed 15 February 2019]

De Sena, A. (ed.) (2015) 'Experiencias hechas cuerpos y emocionalidades configuradas en torno a las políticas sociales: Un abordaje de las políticas sociales desde los Estudios Sociales de los Cuerpos y las Emociones'. Documento de Trabajo No. 5. Centro de Investigaciones y Estudios Sociológicos (CIES). Buenos Aires: ESEditora. Available at: http://estudiosociologicos.org/ [Accessed 15 February 2019].

De Sena, A. (ed.) (2016) *Del Ingreso Universal a las "transferencias condicionadas", itinerarios sinuosos.* Buenos Aires: ESEditora. Available at: http://estudiosociologi cos.org/-descargas/eseditora/del-ingreso-universal-a-transferencias-condicionadas/ del-ingreso-universal-a-transferencias-condicionadas.pdf [Accessed 10 February 2019]

De Sena, A. (2016) 'La ocupabilidad como forma de política social', *Intersticios Revista sociológica de pensamiento crítico*, 10(2), July, pp. 35–49.

De Sena, A. and Cena, R. (2014) '¿Qué son las políticas sociales?. Esbozos de respuestas', in A. De Sena (ed.) *Las políticas hecha cuerpo y lo social devenido emoción: lecturas sociológicas de las políticas sociales.* Buenos Aires: ESEditora/ Universitas. Editorial Científica Universitaria, pp.19–50.

Esping Andersen, G. (1993) *Los tres mundos del Estado del Bienestar.* Valencia: Editions Alfons el Magnánim IVEI.

Galasso, N., Feroci, G., Pfeifer, K. and Walsh, M. (2017) *The Rise of Populism and its Implications for Development NGOs.* Oxfam Research Backgrounder series: www. oxfamamerica.org/riseofpopulism [Accessed 12 March 2019]

Galito, M. (2018) 'Populism as a political phenomenon', *JANUS.NET e-journal of International Relations*, 9(1), May–October. DOI: https://doi.org/10.26619/ 1647-7251.9.1.4 [Accessed 20 March 2018]

Gidron, N. and Bonikowski, B. (2013) 'Varieties of Populism: Literature Review and Research Agenda'. Weatherhead Center for International Affairs, Harvard University, *Working Paper Series, N. 13–0004*, pp. 1–38.

Grassi, E. (2000) 'Procesos Político-culturales en torno del trabajo. Acerca de la problematización de la cuestión social en la década de los 90 y el sentido de las "soluciones" propuestas: un repaso para pensar el futuro ', *Revista Sociedad*, 16. Buenos Aires: Facultad de Ciencias Sociales, UBA, pp. 49–81.

Halperin Weisburd, L. et al. (2008) *Políticas sociales en la Argentina: entre la ciudadanía plena y el asistencialismo focalizado en la contención del pauperismo.* Cuaderno del CEPED No. 10. Buenos Aires: Facultad de Ciencias Económicas, UBA.

Ibarraran, P., Medellín, N., Regalia, F. and Stampini, M. (eds) (2017) *Así funcionan las transferencias condicionadas. Buenas prácticas a 20 años de implementación.* Banco Interamericano de Desarrollo División de Protección Social y Salud. Available at: https://publications.iadb.org/en/publication/how-conditional-cash-transfers-work [Accessed 20 November 2018]

Ianni, O. (1975) *La formación del Estado populista en América Latina.* Mexico: Popular Series Era.

Malmi, A. (2018) 'Los Programas de Transferencia Condicionada de Ingreso en el contexto africano: Un estudio exploratorio de sus impactos en Burkina Faso', in A. De Sena (ed.) *La intervención social en el inicio del siglo XXI: transferencias condicionadas en el orden global.* Buenos Aires: ESEditora, pp. 77–102. Available at: http://estudiosociologicos.org/portal/la-intervencion-social-en-el-inicio-del-siglo-xxi-transferencias-condicionadas-en-el-orden-global/. [Accessed 10 March 2019]

Medellín, P. (2004) *La política de las políticas públicas: propuesta teórica y metodológica para el estudio de las políticas públicas en países de frágil institucionalidad.* Serie: Políticas Sociales, No. 93. Santiago: CEPAL.

Moreno, L. (2013) 'Pobreza y políticas sociales en la Argentina, 1854–1955', *Voces en el Fénix*, 4(23), pp. 7–13. Available at: www.vocesenelfenix.com [Accessed 12 November 2017]

Müller, J. (2017) *The Age of Perplexity: Rethinking the World We Knew: The Rise and Rise of Populism?* Madrid: BBVA, OpenMind, Penguin Random House.

Pastor Seller, E., Sotomayor Morales, E. and Cortés Moreno, J. (2018) 'Situación actual, evolución y tendencias del sistema de rentas en España: las Rentas Mínimas de Inserción', in A. De Sena (ed.) *La intervención social en el inicio del siglo XXI : transferencias condicionadas en el orden global.* Buenos Aires: ESEditora, pp. 103–124. Available at: http://estudiosociologicos.org/portal/la-intervencion-social-en-el-inicio-del-siglo-xxi-transferencias-condicionadas-en-el-orden-global/. [Accessed 12 November 2018]

Scribano, A. and De Sena, A. (2013) 'Los planes de asistencia social en Buenos Aires: una mirada desde las políticas de los cuerpos y las emociones', *Aposta Revista de Ciencias Sociales*, 59. Available at: www.apostadigital.com/index.php. [Accessed 22 November 2018]

Scribano, A. and De Sena, A. (2018) 'Flattened: Social Policies and Politics of Sensibilities', in *Politics and Emotions.* Houston, TX: Studium Press LLC, pp. 78–112.

Scribano, A. and De Sena, A. (2014) 'Prácticas educativas y gestión de las sensibilidades: aprehendiendo a sentir', *Revista Publicatio UEPG: Ciências Humanas, Linguistica, Letras e Artes*, 22, pp. 117–129.

Scribano, A., Huergo, J. and Eynard, M. (2010) 'El hambre como problema colonial: Fantasías Sociales y Regulación de las Sensaciones en la Argentina después del 2001', in A. Scribano and M. Boit (eds) *El purgatorio que no fue: Acciones Profanas entre la esperanza y la soportabilidad.* Buenos Aires: Ed. CICCUS, pp. 23–49.

Scribano, A. and Eynard, M. (2011) 'Hambre individual, subjetivo y social (reflexiones alrededor de las aristas límite del cuerpo)', *Boletín Científico Sapiens Research*, 1(2), pp. 65–69.

Scribano, A. (2004) *Combatiendo Fantasmas.* Santiago: Universidad de Chile, Facultad de Ciencias Sociales.

Scribano, A. (2013) 'Una aproximación conceptual a la moral del disfrute Normalización, consumo y espectáculo'. *Revista Brasileira de Sociologia da Emoção*, 12 (36), Dezembro, pp. 738–750.

Scribano, A. (2012) 'El capitalismo como religión y segregación racializante: dos claves para leer las fronteras de la gestión de las emociones', in I. Picheira Torres (ed.) *Archivos de Frontera: El gobierno de las emociones en Argentina y Chile del presente.* Santiago: Editorial Escaparate.

Scribano, A. (2008) 'Sensaciones, conflicto y cuerpo en Argentina después del 2001', *Espacio Abierto*, 17(2), pp. 205–230.

Scribano, A. (2013) 'La religión neo-colonial como la forma actual de la economía política de la moral', *Prácticas y Discursos*, 2(2) pp. 1–20.

Soares, S. (2012) 'Bolsa Família, Its Design, Its and Possibilities for the Future'. Working Paper No. 89. Brazil: International Policy Center for Inclusive Growth. Poverty Practice, Bureau for Development Policy, UNDP.

Stampini, M. and Tornarolli, L. (2012) 'The Growth of Conditional Cash Transfers in Latin America and the Caribbean: Did They Go Too Far?' Inter-American Development Bank. www.iadb.org/ [Accessed 22 November 2018]

Ştefănel, A. (2016) 'Notes on Populis', *Revue Roumaine de Philosophie*, 60(1), pp. 141–149.

Website

FAO: www.fao.org/3/y4940s/y4940s08.htm [Accessed 23 March 2019]

7 Losing the battle to take back control?

Clashing conceptions of democracy in the debate about Brexit

Thomas Jeffrey Miley

Introduction

Throughout the long, tombstone grey winter and into the early, more occasionally blue-skied spring of 2019, the debate over Brexit thoroughly dominated the political agenda in "Old Blighty", aka not-so-Great Britain. The subject came close to an obsession, literally speaking, i.e. in the clinical sense of the term. The prime minister's felicitous phrase that "Brexit means Brexit", was rendered by now most infamous, an empty signifier if ever there was. Her days in 10 Downing Street, were most certainly numbered; her hold on power, ever-more precarious; her oh-so-hard-bargained "Goldilocks" deal for an orderly withdrawal from the European Union, "not too hard, not too soft", thrice rejected by the House of Commons.

With the prospect of a so-called "no-deal Brexit" looming larger by the day, and with the original deadline for withdrawal, 29 March, quickly approaching, in what was dubbed by Brexit hardliners as nothing short of a constitutional revolution, the House of Commons sought to seize control of the process from May's government by holding a series of indicative votes. The outcome of which, to the shock and dismay of many throughout the Isles, and the amazement of almost all observers across the Channel, turned out to be the rejection of all the alternatives put up for a vote, on not just one but two separate occasions. What they call gridlock, across the Atlantic.

There are multiple, contending, contradictory, to some extent even incommensurate discourses about the significance of Brexit. On the one side, there is a discourse that emphasises xenophobia and racism, propagated in the academy by the likes of Nadine El-Adny (2016), Guminder Bhambra (2017) and Satnam Virdee (2017), who "locate Brexit in the pragmatics of race, citizenship, and Empire", viewing Brexit as reflecting a public mood of post-colonial *melancholia* (Gilroy 2006), and who thereby apply a hermeneutic of suspicion, interpreting the dynamics surrounding Brexit as driven fundamentally by a deeply ingrained if resurgent bout of "xeno-racism" (Fekete, 2001). On the other side, there is a discourse that stresses instead sovereignty and democracy, propagated in the academy by the likes of Chris Bickerton (2017), David Lane (2016) or Mark Ramsden

(2018), who insist that Brexit should be interpreted as first and foremost a people's revolt against the undemocratic, bureaucratic and neoliberal straitjacket that is the European Union.

Each of these discourses undoubtedly captures part of the truth. In fact, one of the principal "virtues" of the Leave campaign's most memorable slogan, "Take Back Control" (at least insofar as campaign slogans can be said to possess "virtues" at all), was its success in appealing simultaneously to the xeno-racist and the democratic impulses of the populous (at least insofar as these two impulses can be treated as separate at all). An appeal to the dark side of democracy, perhaps.

Brexit and "xeno-racism"

The interpellation of the xeno-racist impulses of the populace over the course of the referendum campaign was most blatantly performed by the man who arguably did the most to set the agenda, the United Kingdom Independence Party's Nigel Farage. Figure 7.1 shows how an image is worth a thousand words.

Further buttressing the plausibility of the discourse of xeno-racism is the fact that the far right seemed so emboldened by the Brexit victory – as evidenced most dramatically by the unprecedented spike in religious and racially motivated hate crimes that took place in the wake of the referendum result (Bulman, 2017; Devine, 2019).

Finally, there is the evidence that can be culled from the copious public opinion survey-based research that has proliferated on the subject of Brexit.

Figure 7.1 Breaking point

The people who run the British Social Attitudes Survey concluded that the issue of immigration was "at the heart" of the Brexit vote, with fully 73 per cent of those who registered concerns about immigration voting in favour of Brexit, compared with 36 per cent who did not register any such concerns (BSA #34, 2016).

Attitudes about immigration are themselves endogenous to broader processes, both political and social. They are "mediated", so to speak. And in Britain, the tabloid press has played an especially nefarious role in propagating a toxic ideological climate, manufacturing a comic-book consciousness of sorts, one in which the migrant-cum-refugee appears as a familiar, stock character, portrayed as a villain upon whom public fears and anxieties can be conveniently projected. Day in and day out, British people are inundated with images and headlines encouraging them to view migrants in almost exclusively negative terms. The tabloid press repeatedly "informs" people: (1) that there are too many migrants, that they come in "floods", "swarms" and "hordes"; (2) that migrants are criminals, portrayed as "illegals, as "thieves", even "murderers"; (3) that migrants are stealing "our" jobs; and (4) that when they are not stealing "our" jobs, they are taking advantage of "our" benefits, indeed that they are "benefits scroungers" (Walsh, 2018). No wonder the moral panic surrounding the subject.

Furthermore, survey research compellingly shows that those most susceptible to being successfully interpellated by such negative depictions of migrants – that is, those who come to accept such depictions at face value – also tend to prove susceptible to being successively interpellated by a whole host of other right-wing, reactionary appeals. So much so that registering concern about immigration is often taken as one of the main indicators of possessing an "authoritarian" value orientation. And in fact, according to the British Social Attitudes surveys, among those categorised as possessing such "authoritarian" values, fully 72 per cent registered support for Brexit, compared with a mere 21 per cent of those categorised as possessing "libertarian" values (BSA #34, 2016).

If we break it down further, we find, for example, that whereas only 29 per cent of those who register favourable attitudes towards multiculturalism supported Leave, fully 81 per cent of those with negative attitudes towards multiculturalism did. Likewise, whereas approximately 40 per cent of those who register favourable attitudes towards feminism supported Leave, among those who have negative attitudes towards feminism, support for Leave rose to 74 per cent. Similarly, among those with favourable attitudes towards the Green movement, support for Brexit was a mere 38 per cent, compared with fully 78 per cent of those who reject the Green movement (Lord Aschcroft, 2016).

Brexit by the numbers

In sum, evidence from survey research would seem to validate the discourse of those who emphasise xeno-racism as a crucial hermeneutic key for understanding Brexit. But this is only part of the story revealed by survey research. For such research also shows that attitudes towards Brexit vary quite significantly across cohorts divided by age, educational levels, class, partisan preferences, racial identity, religious identity, national identity and region as well.

The generational factor was particularly stark – with close to three-quarters of those between 18 and 24 preferring Remain, compared with only 40 per cent of those over 65 (Lord Ashcroft, 2016). So too did educational level correlate quite strongly with preferences – with less than one in three of those with no qualifications opting for Remain, compared with approximately two-thirds of those with university degrees. Somewhat less stark, though still clear, were the differences among people belonging to different class categories – with close to six in ten of those belonging to the highest class category (AB) preferring Remain, compared with approximately four in ten of those belonging to the lowest class category (DE) (Lord Ashcroft 2016).

In terms of partisan preferences, just over 40 per cent of Conservative voters opted for Remain, compared with close to two-thirds of Labour voters. Race mattered too, with just under half of those who identify as white opting for Remain, compared with seven in ten of those who identify as "BAME" (black and minority ethnicity). Likewise religion, with only four in ten of those describing themselves as Christians opting for Remain, compared with nearly seven in ten of those who describe themselves as Muslims (Lord Ashcroft, 2016).

So too with region. In Scotland, over six in ten preferred Remain, and in Northern Ireland there was a clear majority (55 per cent) of "Remainers" as well. A similarly clear majority of Londoners (55 per cent) opted for Remain (Lord Ashcroft, 2016).

In terms of national identification, the stronger the sense of Englishness one registers vis-à-vis Britishness, the more likely one was to prefer Brexit. While less than four in ten of those who identify as British, not English, opted for Brexit, the proportion rose to over seven in ten of those who identify as English, not British (Lord Ashcroft, 2016).

Such are the relations revealed by the surveys. The numbers provide us with a map of sorts – allowing us to glimpse how attitudes (and reported voting behaviour), in this case, about Brexit, are embedded within broader constellations of material and social power relations. However, numbers alone are ultimately limited in their ability to help us adjudicate among contending interpretations about the significance of any complex social phenomenon. Sometimes a picture can be worth not only a thousand words, but also a thousand cross-tabs, perhaps even a thousand regressions (see Figure 7.2).

The debate in the House of Commons

But let us leave the numbers, and even the pictures, behind us – for the moment, at least – so as return to the domain of discourse, to the world of words. Not just any words either, but words spoken in Westminster, by the elected representatives of the British body politic, in the House of Commons, on the day that that House sought effectively to seize control of the Brexit

Figure 7.2 Anti-Brexit sentiment at the Irish border

process from Theresa May's government, by forging ahead, over the government's vehement objections, with substantive deliberations followed by a series of indicative votes on an ample set of alternatives to the prime minister's "deal", i.e. her government's proposed terms of withdrawal from the EU.

This unprecedented debate in the Commons took place some two weeks after the House had rejected the government's terms of withdrawal for a second time, thereby forcing the government to apply for an extension to the original Brexit deadline, and less than a week after an estimated million people had marched in London to demand a second referendum, or "people's vote". The terms of the debate reveal a whole lot about the multiple, conflicting, in part contradictory, conceptions of democracy espoused by the country's elected representatives.

One of the more striking features of the political discussions surrounding Brexit in the run-up to the referendum was the relative neglect of the series of serious constitutional quandaries that a yes-vote could foreseeably cause. The then-Prime Minister David Cameron had made a solemn promise to respect the result of the referendum, despite the fact that it was, technically, a non-binding, consultative poll; and in the wake of its outcome, both the House of Commons and the House of Lords long remained nearly unanimous in their deference to the "will of the people" as expressed in that consultation-cum-plebiscite. There had been, as well, virtually no discussion at all about whether there should be a threshold, in terms of levels of participation and/or percentage of the vote, for constituting a clear majority, either before or after the fairly close 52 per cent to 48 per cent outcome in favour of Leave. Nor, for that matter, were any of the thorny issues related to devolution in Scotland or Northern Ireland even really considered until they arose after the vote, as near *fait accompli* – issues such as whether a clear majority for Remain in either of those devolved territories should be recognised and respected, not to mention what to do about the Good Friday Accord's guarantee of a soft border with the Republic of Ireland.

On all such important constitutional questions, the UK's elected representatives had exercised precious little in the way of foresight, deliberation or judgement. And so, the decision by the House of Commons to try to seize control of the process from the government, at the eleventh hour, came as something of a surprise to most who had been watching it defer and demur for the better part of three years. Better late than never, as many of the Parliamentarians themselves put the point on that fateful day; nothing like a quickly approaching deadline to focus the mind.

A crisis of democratic leadership

The referendum had been a gamble by the former bad boy from Eton, Cameron, who had hoped to resolve the long-standing strife in the Conservative Party over the issue of EU membership, to defeat the Euro-sceptic wing of his party once and for all, by appealing directly to "the people". But his bet backfired badly, in no small part due to the demagogic antics of fellow Etonian and fellow Tory Boris Johnson. A spat between two "public" (sic) school boys, camouflaged as a bout between the establishment and the people, was bound to become a constitutional crisis of the highest order.

Atrocious leadership, to say the least. Which brings us back to the subject of democracy. The term democracy means, literally, rule (*"kratos"*) by the people (*"demos"*). Nowadays, representative government is often equated with democracy, despite the fact that, as Bernard Manin has reminded, it was originally "conceived in explicit opposition to democracy" (Manin 1997, p. 236). The so-called "principal–agent" problem perpetually hinders the democratic credentials of representative institutions. As Rousseau put the point provocatively but poignantly in his *Social Contract*, the sovereignty of the people is inalienable. Therefore, though the people of England (sic) may "regard itself as free", it is in fact "grossly mistaken" – "it is free only during the election of members of parliament. As soon as they are elected, slavery overtakes it, and it is nothing" (Rousseau [1762]/2012).

This is because, in Britain, as Professor Nicholas Boyle, fellow at Magdalene College in Cambridge, felt compelled to chastise the *Financial Times* for having forgotten in its exasperated editorial penned in the wake of the referendum, that "in the UK the people are not sovereign". Indeed, according to the British Constitution, "the sovereign is the Crown in Parliament" (2016).

The *Financial Times* can perhaps be forgiven for having seemingly forgotten the terms of the British Constitution, since the Members of the House of Commons at least pretended to as well. Ironically enough, since, as Professor Boyle also emphasises in his letter to the editors of the *Financial Times*, one of the main arguments advanced by the Leave campaign had been that no extra-parliamentary authority (save the Crown) can constrain the sovereignty of the Parliament. What goes for foreign powers should go for referenda as well, so Boyle pithily concludes (2016). Though such impeccable constitutional logic seemed lost upon the country's elected representatives, not to mention its chattering classes, at least up until the eleventh hour. Evidently, what had once been a matter of pride openly asserted by the ruling class in Britain appears to have been reduced to a dirty little secret, mentioned without embarrassment only by a select few who inhabit the exclusive confines of the colleges of Oxbridge.

Could it be that the democratic ideal had at last conquered even the great bastion of representative government, Great Britain? Self-determination, the will of the people, as expressed directly in a plebiscite, now recognised by its ruling class, its representatives, as trumping even the sovereignty of the Crown in Parliament? Edmund Burke must be rolling over in his grave!

Another feature of the public discussion surrounding Brexit is the common conflation of the instrument of a referendum with direct democracy. However, a referendum is qualitatively different from direct democracy, in at least two crucial respects, since the latter requires direct citizen participation in the process of deliberation and debate, and most often aims at achieving a general consensus, if not unanimity, among those who deliberate. In a referendum, by contrast, no direct participation by the citizenry in the process of deliberation is required; and a majority verdict, not consensus, much less unanimity, is sought. Participation is, rather, reduced to the act of choosing among pre-selected, usually binary, alternatives.

What referenda and direct democracy do share in common, however, is their reduced reliance on the role of representatives. Which renders both instruments closer to the democratic ideal, at least as this ideal is most frequently conceived nowadays – for example, as defined by the dean of twentieth-century democratic theory Robert Dahl, on page one of his contemporary classic *Polyarchy*, in terms of "the continuing responsiveness of the government to the preferences of its citizens" (1971).

Even so, as Juan Linz has lamented, this near-ubiquitous emphasis on responsiveness in contemporary definitions of democracy has entailed an equally near-ubiquitous neglect of consideration of what democratic leadership should look like (2012). So much so that the very term "democratic leadership" comes across as an oxymoron to many, if not most, people these days. Thus, perhaps, the hesitancy of the Members of Parliament to lead at all, especially if and when it might mean contradicting the will of the people as allegedly expressed directly in a referendum.

But lest Burke's defence of the legitimate role of representatives in matters pertaining to rule be also forgotten, let us remind the reader of his eloquent appeal to the Bristol electorate, in which he famously insisted:

> My worthy colleague says, his will ought to be subservient to yours. If that be all, the thing is innocent. If government were a matter of will upon any side, yours, without question, ought to be superior. But government and

legislation are matters of reason and judgment, and not of inclination; and what sort of reason is that in which the determination precedes the discussion, in which one set of men deliberate and another decide, and where those who form the conclusions are perhaps three hundred miles distant from those who hear the arguments?

To deliver an opinion is the right of all men; that of constituents is a weighty and respectable opinion, which a representative ought always to rejoice to hear, and which he ought always most seriously to consider. But *authoritative* instructions, *mandates* issued, which the member is bound blindly and implicitly to obey, to vote, and to argue for, though contrary to the clearest conviction of his judgment and conscience, – these are things utterly unknown to the laws of this land, and which arise from a fundamental mistake of the whole order and tenor of our Constitution.

Parliament is not a *congress* of ambassadors from different and hostile interests, which interests each must maintain, as an agent and advocate, against other agents and advocates; but Parliament is a *deliberative* assembly of *one* nation, with *one* interest, that of the whole – where not local purposes, local prejudices ought to guide, but the general good, resulting from the general reason of the whole ([1774]/1999, pp. 155–157).

The responsibility of the representative to rule in accordance with his or her judgement, and in accordance with his or her conscience, about what constitutes the interest of the whole polity, even above and beyond the opinions of his or her constituency, once understood and considered an honour and a fundamental element of the British Constitution, is only rarely invoked, at least openly, in this democratic age. Indeed, the "delegate" model, or "imperative mandate", would seem to have decisively vanquished any such notion of a "free mandate", at least in dominant articulations of the proper role of elected representatives in the process of rule in Britain, as we shall shortly see.

The duty to respect the will of the people

During the historic debate about Brexit that took place in the House of Commons on 27 March, the argument aired perhaps most frequently and, among staunch Brexiteers, most vehemently, had to do with the need to respect the result of the referendum. Over and over, among supporters of Brexit from across the partisan spectrum the point was emphasised, that, in the words of Conservative representative from the constituency of Mid Norfolk, George Freeman, "we have to honour and respect the referendum result" (Hansard 2019, Vol. 657). As the Conservative from Carlisle, John Stevenson, put the point: "My starting point is very simple: this country voted to leave the EU. I therefore firmly believe that we must leave the EU institutions." Or, in a more polemical articulation from the Labour "Leaver" from Vauxhall, Kate Hoey, who, after expressing sadness about the debate even taking place, added:

The one group of people we cannot blame, however, are the people of this country who in the referendum voted to leave, thought they would be listened to and were told by everyone, including the former Prime Minister, that their vote mattered and would be implemented, whatever that decision.

According to this view, the elected representatives in the House of Commons have an obligation, a mandate, to implement the decision of the British people, as expressed in the referendum, to leave the European Union. Any other course of action would amount to nothing less than an affront to democracy. Some would add to such a direct mandate from the people an additional obligation to honour the electoral manifestoes of both the Conservative and Labour parties on which members of these parties were elected in the 2017 election, both of which included clauses promising to respect the 2016 referendum outcome. As the Conservative representative from Yeovil, Marcus Fysh, argued:

> [L]et us actually remember the people in all this. They voted two years ago to leave the European Union and then they voted in an election in which we stood on a manifesto saying that we would leave the European Union and its two main pillars, the single market and the customs union, which are integral to what the European Union is. They want their instruction to be carried out now.

In a similar vein, Labour's Katie Hoey referred to a letter she received from a member of her constituency, along the following lines:

> A constituent wrote to me saying that he had thought that the manifestos of the Labour party and Conservative party – the two main parties – had said, "We will implement the result of the referendum." There is nothing difficult about the word "leave". It is very simple. Members have deliberately made it difficult here.
> My constituent wrote:
> "Can we the electorate now expect that anything promised in a manifesto is to be honoured, that it should be written into law, that, if you promise a course of action, you must follow through and make it happen?"

To which she added, in no uncertain terms: "I think that we are in a very dangerous situation in the House. We are trying to thwart the will of the people, but democracy cannot be compromised."

Some of the representatives, furthermore, made it clear that their sense of obligation to the mandate they had received from the British people was strong enough to override their own personal preferences or judgement. The Conservative representative from Harlow, Robert Halfon, expressed such a view rather succinctly: "I passionately believe that we have to follow the 2016 referendum result, even though I voted remain." Some mentioned promises they had made to their constituency in this regard, such as the Conservative from South Suffolk, James Cartlidge, who insisted: "I campaigned to remain, but I promised my constituents that I would accept the result of the referendum that my colleagues and I voted into law."

Likewise, Halfon's and Cartlidge's party mate George Freeman prefaced his remarks about the need to honour and respect the referendum result by pointing out that he believed so despite the fact that "I was a remain Minister in the last Government". Freeman went on to add a further criterion of responsiveness to which he felt bound – namely, the will of his constituency. In his words: "I have been very clear that we have to honour and respect the referendum result both nationally, in my duty as a Member of this House, and locally, in my responsibility and duty to the people of Mid Norfolk, who voted 62% to leave."

On which alleged duty – to respect the will of the "nation", or to respect the will of the local constituency – should trump in case of a conflict between the two, Freeman could afford to stay silent. Though the matter is not necessarily merely an academic one, especially given that recent polls showing a rather solid majority has emerged in favour of remain these days – a majority of somewhere between 8 and 12 per cent, apparently – though the proportion of local constituencies that voted Leave remains close to two-thirds (64 per cent).

The duty to respect the will(s) of the people(s) of Scotland and Northern Ireland

Representative Freeman's mention of the will of the "nation", not merely the will of "the people", brings up another point – the constitutional ramifications of devolution – largely ignored by English members of all parties in Parliament. In a word, what about the will of the Scottish nation or, for that matter, Northern Ireland? Can the clear majorities in favour of Remain that were registered in both of these devolved constituencies in the referendum be so blatantly, ignored? If so, what does such treatment portend for the future of the Union?

As the Scottish National Party representative for Ross, Skye and Lochaber, Ian Blackford, complained:

> Scotland did not ask for this crisis; nobody asked for this chaos. Of course, Scotland voted to remain in the EU. We voted overwhelmingly to protect our economy and the freedoms and the values that the European Union gives to the people of Scotland. Scotland is a European country; historically, we have been a European country. Economically, socially and culturally, we benefit from our membership.
>
> Today the SNP laid a motion to ensure that Scotland's voice is heard, because Scotland's wishes have been completely ignored during the Brexit process ...

It is most certainly not the partnership of equals that the Prime Minister had promised us. It is one where we are told, quite simply, that our votes do not count, where we can be stripped of our European citizenship – and for what? – and where we will pay a price economically, socially and culturally ...

Likewise with Northern Ireland. When, at the outset of the debate, the Conservative from Basildon and Billericay, John Baron, was urging the House to favour "no deal" as an option, he was interrupted by the Independent Anna Soubry from Broxtowe, who interjected rather emphatically:

Let me take a moment to remind the House and in particular the hon. Gentleman that Northern Ireland has not had a Government since January 2017. We have no Ministers in Northern Ireland. The head of the Northern Ireland civil service has warned as recently as the beginning of this month of the 'grave' consequences for Northern Ireland if we were to leave without a deal. Does the hon. Gentleman have any respect at all for the head of the civil service in Northern Ireland or indeed for the people of Northern Ireland?

To which point Mr Baron most tellingly replied: "Having served there in the 1980s and got the medals to prove it, I take into account what the people of Northern Ireland, as part of our Union, have to say. At the same time, we are part of a United Kingdom." The limits to devolution could hardly be expressed more clearly – shared sovereignty it is not. The will of the whole trumps the will of the part, regardless of the consequences for the part. Good Friday Accord or not.

If staunch supporters of Brexit were quick to refer and defer to the will of the people as allegedly expressed in and emanating from the result of the referendum, for their part "Remainers" in the House did not refrain from calling the status and even the fairness of the referendum into question. The Labour representative from Brentford and Isleworth, Ruth Cadbury, remarked in such a vein: "The referendum was advisory – a simple yes or no – with little information and many lies." To underscore the point, she continued: "Many people challenge me on the manifesto phrase about respecting the results of the referendum. Well I do respect the reasons why most people who voted leave did so – because of the lies ..."

Along even more polemical lines, the Labour representative from Cambridge, Daniel Zeichner, who labelled himself "a passionate Remainer", alluded to the prospect of Russian interference in the referendum campaign, by making reference to "Russian bots crawling all over parts of one of the campaigns".

The right to a second referendum

However, a more frequent rhetorical tactic on the part of the staunch "Remainers" was not to deny that the will of the people was expressed in the

2016 referendum, but instead to suggest that the will had changed since then, or at least that the will of the people needs to be gauged again. In such a vein, the Independent from Totnes, Sarah Wollaston, who resigned from the Tories in February of 2019, interjected to query her former party mate, Nick Boles, while he was expounding upon virtues of his Common Market 2.0 proposal, to put to him the question: "Would it not be best ... at least to check that it has the consent of the people? Would he agree to link it to a public vote, so that we can check that it really is the will of the people?"

More directly still, the Labour representative from Holborn and St Pancras, shadow Brexit secretary Keir Starmer, insisted: "[I]t is now clear that any Brexit deal agreed in this Parliament needs further democratic approval." Likewise, fellow Labour representative from Derby South, Margaret Beckett, asked: "But how, in all conscience, can we alone in this House force through such a decision on [the people's] behalf without allowing them any say as to whether that is still their view?" And she proceeded to insist: "As with the Good Friday Agreement, the outcome should go back to the people for confirmation."

Some in the House went so far as to formulate a second referendum in terms of the right of the people to change their minds. Along these lines, the Labour representative from Leeds Central, Hillary Benn, argued: "The Prime Minister said 108 times that we would definitely leave on 29 March, but she changed her mind and we are not. Why is it that the only people in this debate apparently not allowed to change their minds are the British people? How democratic is that?" Or even straighter to the point, in the words of the Independent from Broxtowe, Anna Soubry:

> What a profound irony, and some would say, a disgrace, verging on hypocrisy. Hon. and right hon. Members will expect and enjoy the right to change their minds and their vote, but not allow the people of this country the same right. That is why I shall be supporting the motion to allow whatever we agree and decide on to go back to the British people. They are entitled also to change their minds and their votes, especially when they see, whatever way you cut it, Brexit will make our country worse off.

Along similar lines, the Independent from Streatham, Chuka Umunna, who resigned from the Labour Party in February 2019, emphatically insisted:

> Let us not forget that that referendum was held three years ago, when 37% of registered electors voted to leave. The most recent poll of the British people was held in 2017, when the Conservative hard Brexit was put to the British people and the party of Government lost its majority. If that were not the case, we would not be having this protracted process right now. Above all, I say to those who talk about the will of the people that democracy is not static; it is a dynamic thing. We in this country did not choose to have a system in which we have one general election and a one-party state and in which we never go

back to the people for their view on things as our country and the world change and adapt.

Others emphasised, as Beckett had suggested above, that the British people have the right to the *last word* in the process. Along such lines, the Labour representative from Wirral South, Allison McGovern, employed the language of ratification. In her words: "I think the only thing left is to find a reasonable, tolerable and acceptable form of Brexit and ask for it to be ratified by the British public." Or, in the words of the Labour representative from Stockton South, Paul Williams: "There is nothing threatening to democracy about testing the public's opinion ... Brexit is far more complicated than anybody expected and we now have a duty to bring the public back into our discussions as we reach this vital, difficult stage in the process."

If the people had the first word in this process, so too should they have the last word, so the advocates of a second referendum insisted. In the words of the Labour representative from Sedgefield, Phil Wilson: "[T]he British people have the right to the final say on this country's future direction."

Some making such an argument admitted that a second referendum is bound to be controversial. The Labour representative from Newcastle-under-Lyme in this vein acknowledged: "I certainly dread the thought of a second referendum. Powerful, loud and deep-pocketed voices tried to drown out debate with cries of 'Betrayal'." Even so, he insisted: "[W]e have to be brave. In the interests of our country, we should not shy away from giving the people, including young citizens who are 16, a final say on their future."

A final say for the people as the brave and democratic thing to do, so the argument goes. Clashing conceptions of what democracy requires, for sure, given that the Conservative MEP from Bosworth, David Treddinick, could refer to the prospect of a second referendum as nothing less than "the end of democracy". On this thorny matter of the dictates of democracy, some of the advocates of a second referendum grabbed the bull by the horns. Such as the Labour representative from Barnsley East, Stephanie Peacock, who thus expounded:

> Some say that what we are promising is undemocratic because the people have already had their say. Yes, they have. But they did not have a say on the current Brexit deal – or, in fact, on any Brexit deal – and they should. When I suggest that the electorate should be given the final say on what the deal should be, some people react as if the only ones who would be allowed to vote are those who voted to remain. People should have the right to changes their minds – not just from leave to remain, but from remain to leave. I do not believe that MPs in this House today, who are elected, in theory, for five-year terms, should have the final say on an issue that will affect our electors, and their families and descendants, for years to come. If that were to happen, it would not reflect well on the establishment, however it is appointed or elected.

The final say should not be given to Members of this House exclusively. The final say belongs to the people. Brexit started with the people and it should end with the people.

To which point, the Labour representative from Vauxhall, Kate Hoey, indignantly replied: "The one thing that must not happen today is the people of this United Kingdom being told, 'You were too stupid, racist or ignorant to vote the right way, and now we want you to vote again in a separate referendum, because we think you might have changed your mind'."

In addition to survey evidence from which "Remainer" representatives appear to have glimpsed an emergent will of the people, a few made reference to evidence from recent mobilisations as well. In such a vein, Ian Blackford, the SNP MP from Ross, Skye, and Lochaber, would mention that "[o]n Saturday, more than one million people marched to ask that they get the chance to vote on their future within the European Union." Likewise, Gareth Thomas, the Labour representative from Harrow West, even placed himself among the million marchers, saying: "I was proud to be one of one million who marched in London on Saturday." While Wera Hobhouse, the Liberal Democratic MEP from Bath, estimated the number of marchers to be even higher, and went so far as to suggest that they constitute "the true people". In her words:

Between 1 million and 2 million people marched peacefully in the streets of London – young and old, from all backgrounds, from different political parties and none. Do they not count? Are they not the real British people, determined but polite? Does Parliament listen to people only when they throw stones or send us death threats?

Such mentions of the million-person march for a so-called "people's vote" provoked a sardonic reply from the Conservative MEP from Brigg and Goole, Andrew Percy, who quipped: "The only march that I am interested in is the march of my constituents to vote in the 2016 referendum." Contradictory mandates, to say the least.

The duty to lead

If both staunch "Brexiteers" and staunch "Remainers" were quick to invoke the will of the people, expressing a sense of obligation, a duty to follow – to be responsive to – a clear mandate emanating from the people, at least as they interpreted it, there was much more reticence to invoke a sense of responsibility, a duty to lead, to act in accordance with the judgement and conscience of honourable and right honourable members of the House of Commons. Though such expressions were not entirely absent from the debate either.

The Independent from Streatham, Chuka Umunna, who defected from the Labour Party in February 2019, for example, in attempting a rebuttal of those who seemed complacent about the prospect of a no-deal Brexit, insisted: "No one in this House has a mandate to destroy people's jobs and livelihoods, but we know that a no-deal exit would do that because the Cabinet has produced its own briefing papers telling us that that is a fact." This before invoking a sense of obligation not to any existing constituency, but instead to future generations. In Umunna's words: "This is what is at stake here; this is what we have to think about when we make this decision. This is not about us so much as about future generations, and it is important that we do right by them."

Along similar lines, Ruth Cadbury, the Labour MEP from Brentford and Isleworth, emphasised that she too felt bound to follow first and foremost her conscience, by insisting: "Although my constituency voted to remain, I would probably take the same position even if my constituency was a leave-voting area because of my duty to my country."

The duty to country, to lead rather than to follow, in representative Cadbury's judgement, like that of Umunna, translates into a duty to push for "Remain". Nevertheless, most of the MEPs who invoked a duty to lead did not come down so clearly on one side of the debate. Rather, more often than not, they invoked their sense of responsibility, their duty to lead, in the service of the ideal of forging a compromise. Such as, for example, when the Conservative representative from Rushcliffe, Kenneth Clarke, argued that "[m]y duty now is to exercise my own judgment as to what is in the national interest, [what] will minimise the damaging consequences and will perhaps save some of the better features for future generations", only to link this line of argument to the need to find "some sort of cross-party consensus and some search for a majority that can be sustained through the difficult and long negotiations that will be required to reach agreement on our final relations with the European Union".

This is leadership as the courage to find a compromise, to overcome divisions and reach a middle ground. As articulated perhaps most emphatically by Nick Boles, the Conservative representative from Grantham and Stamford, who within the week resigned from his post as party whip and leave his party over its failure to seek compromise. While urging support for his "Common Market 2.0 proposal", Boles insisted:

> I, too, want to make the case for compromise, not as something cowardly but as something courageous ... Each of us today is a leader. The Prime Minister has one vote, the Leader of the Opposition has one vote, and so does every other right hon. and hon. Member. In years to come, the question that our children and grandchildren will ask us is this: in that historic week when Parliament took charge of the nation's destiny, what did you do? Did you stand up and lead? Did you step forward to help reunite our country, or did you hang back in your party trench waiting to be told what to do and where to go? I have already made my choice at the cost of my future career in this House. It is now time for others to choose. To all right hon. and hon. Members I say this: if you choose Common

Market 2.0 this evening, the history books will record it as the moment that our country turned a corner and the part you played will be something of which you will be forever proud.

Conclusion

Thus, not all of the Members of the House of Commons abdicated the responsibility to act in accordance with their own judgement, their own conscience, about what constitutes the national interest, to lead rather than follow. Indeed, the debate itself was hailed as a turning point precisely because it signified an attempt to seize the Brexit process, by the House of Commons, from an ever-weaker government, an attempt by the honourable and right honourable Members of the country's representative chamber – constitutionally speaking, the seat of sovereignty – to at last lead, to take back control.

However, most tellingly, the House ended that fateful day by voting down all the alternatives on offer to the government's deal, and did so again within the week. The House itself, like the public, was bound to remain divided. Like Melville's fictional scrivener Bartleby, when it came to the prospect of exercising leadership, on the whole, upon second thoughts, the Members of Parliament would prefer not to (Melville [1853]/2014).

References

El-Adny, N. (2016) 'Brexit as Nostalgia for Empire'. Online at: http://criticallegalthinking.com/2016/06/19/brexit-nostalgia-empire/. [Accessed 30 March 2019].

Bhambra, G.K. (2017) 'Locating Brexit in the Pragmatics of Race, Citizenship and Empire', in W. Outhwaite (ed.) *Brexit: Sociological Responses*. London: Anthem Press.

Bickerton, C. and Tuck, R. (2017). 'A Brexit Proposal'. Online at: https://thecurrentmoment.files.wordpress.com/2017/11/brexit-proposal-20-nov-final1.pdf. [Accessed 30 March 2019].

Boyle, N. (2016) 'In the UK, the Sovereign is the Crown in Parliament', *Financial Times*, 28 June. Online at: https://www.ft.com/content/1eeaa1f8-3a26-11e6-a780-b48ed7b6126f. [Accessed 30 March 2019].

British Social Attitudes #34 (2016). Online at: www.bsa.natcen.ac.uk/media/39149/bsa34_brexit_final.pdf. [Accessed 30 March 2019].

Bulman, M. (2017). 'Brexit Vote Sees Highest Spike in Religious and Racial Crimes Ever Recorded', *Independent*, 7 July. Online at: https://www.independent.co.uk/news/uk/home-news/racist-hate-crimes-surge-to-record-high-after-brexit-vote-new-figures-reveal-a7829551.html. [Accessed 30 March 2019].

Burke, E. ([1774]/1999) 'Speech to the Electorate of Bristol', in I. Kramnick (ed.), *The Portable Edmund Burke*. New York: Penguin Classics.

Carson, N. (2018) 'Negotiations Are Continuing on How to Avoid a Hard Border', *Belfast Telegraph*, 13 November. Online at: https://www.belfasttelegraph.co.uk/news/brexit/northern-ireland-rejects-hard-border-and-62-say-united-ireland-more-likely-after-brexit-37521930.html. [Accessed 30 March 2019].

Dahl, R.A. (1971) *Polyarchy: Participation and Opposition.* New Haven, CT: Yale University Press.

Devine, D. (2019) 'Hate Crime Did Spike after the Referendum – Even Allowing for Other Factors'. Online at: https://blogs.lse.ac.uk/brexit/2018/03/19/hate-crime-did-sp ike-after-the-referendum-even-allowing-for-other-factors/. [Accessed 30 March 2019].

Fekete, L. (2001) 'The Emergence of Xeno-Racism', *Race and Class* 42(2), pp. 23–40.

Gilroy, P. (2006) *Post-Colonial Melancholia.* New York: Columbia University Press.

Hansard (2019) 'EU: Withdrawal and Future Relationship (Motions)', Volume 657. Online at https://hansard.parliament.uk/commons/2019-03-27/debates/45525049-637 A-47BF-90CA-AF6A2D9F16ED/EUWithdrawalAndFutureRelationship(Motions). [Accessed 30 March 2019].

Lane, D. and Miley, J. (2016) 'Brexit Debate'. Online at: http://archive.cambridge-tv. co.uk/brexit-debate/. [Accessed 30 March 2019].

Linz, J.J. and Miley, T.J. (2012) 'Cautionary and Unorthodox Thoughts about Democracy Today', in D. Chalmers and S. Mainwaring (eds) *Institutions and Democracy: Essays in Honor of Alfred Stepan.* South Bend, IN: University of Notre Dame Press, pp. 227–252.

Lord Ashcroft (2016) 'EU Referendum: "How Did You Vote?" Poll'. Online at: http s://lordashcroftpolls.com/wp-content/uploads/2016/06/How-the-UK-voted-Full-ta bles-1.pdf. [Accessed 30 March 2019].

Manin, B. (1997) *The Principles of Representative Government.* Cambridge: Cambridge University Press.

Melville, H. ([1853]/2014) 'Bartleby the Scivener. A Story of Wall Street', in *I and My Chimney and Bartleby the Scrivener. A Story of Wall Street.* Somerville, TN: Bottom of the Hill Publishing.

Miley, J. and Ramsden, M. (2018) 'Brexit: Arguments For and Against', Debate in *Soc 12: Social Problems in Britain,* University of Cambridge, Lent Term.

Rousseau, J.J. ([1762]/2012) 'On the Social Contract', in D.A. Cress and D. Wootton (eds) *The Basic Political Writings,* Second Edition. Indianapolis, IN: Hackett Publishing Company.

Thomas, M. (2016) 'Nigel Farage with the Poster', *Guardian,* 16 June. Online at: http s://www.theguardian.com/politics/2016/jun/16/nigel-farage-defends-ukip-breaking-p oint-poster-queue-of-migrants. [Accessed on 30 March 2019].

Virdee, S., and McGeever, B. (2017) 'Racism, Crisis, Brexit', *Ethnic and Racial Studies,* 41(10), pp. 1–18.

Walsh, P.W. (2018) 'The Politics of Migration in Britain', Lecture in *Soc 12: Social Problems in Britain,* University of Cambridge, Lent Term.

8   Populism and neoliberalism in Chile

Freddy Timmermann

Populism in populisms

As Torcuato Di Tella et al. (2008, p. 567) state, we must distinguish between populism as a historical subject and populism as a characteristic that can be possessed partially by a large number of political phenomena. Dockendorff and Kaiser (2017, pp. 75–100) argue that it has no ideological coherence or conceptual precision. For Fassin (2018), it is characteristic of each historical development and possesses its particular features that hinder the development of knowledge about it. In Latin America, the main historical expressions of populism appear in the first two-thirds of the twentieth century – with Cardenas in Mexico, Perón in Argentina and Vargas in Brazil as major paradigms – when traditional society began to give up its power spaces in the midst of a capitalist transformation towards industrialisation to replace imports, with popular masses arriving and inserting themselves into politics, generating social inclusion projects with certain perspectives of nationalist movements. It was intended to achieve economic growth with social justice. According to Drake, its main features are to constitute a political movement with an identifiable style of mobilisation, leadership, campaigns and propaganda, which seeks to generate an immediate gratification of mass needs, through personalism, paternalism and nationalism, structuring a heterogeneous coalition oriented to the working-class interests, but which also includes and is directed by medium and high levels sectors, giving birth to an eclectic set of policies adopted in periods of "modernisation", with programmes of national integration that respond to underdevelopment issues.[1] A central aspect observed in the aforementioned historical developments is an expansion of public spending in the implementation of its programmes.

The perspectives on populism are broad, although they are situated, with variations, in the aforementioned aspects, with emphasis on structural functionalist elements in the work of Gino Germani and on class struggle in the writings of Octavio Ianni. Mackinnon and Petrone (2010, pp. 11–55) characterise populism from the existence of a crisis as an emergency condition, the experience of participation as popular mobilisation and its ambiguous character. Giner et al. (1998) think that it has an openly emotional, Manichean and self-assertive rhetoric around the idea of the people, to whom are attributed virtues of justice and political morality. To guarantee the fulfilment of popular desires is more to build an affective bond with an honest and charismatic leader, more than a programme or planned tactics. For Cortés and Pelfini (2017), populism is a concept that during much of the twentieth century allowed us to explain and understand complex political processes of incorporation of broad social sectors that have been neglected in political life, and is currently used by an important part of the social sciences, by the press and by political actors as an indistinct synonym of demagoguery. Enrique Dussel describes the existence of a "pseudo-populism" as a pejorative epithet, constituting a conservative political critique of any measure or social or political movement that is opposed to the trend of globalisation as described by the basic theory of the "Washington consensus". Therefore, he adds, "the populists are always other, anomalous and incurable, which cynically persist in mixing extravagantly, and for their own benefit, the political and social spheres that neoliberalism and Institutionalism had tried to separate forever". His elitist use reproduces the logic of the existence of two lefts: a "good" left closer to the British Third Way, concerned with macroeconomic balances, with the realisation of reforms as far as possible, strictly respecting national institutions, opposed to the "wrong" left which would be post-neoliberal, with refoundational projects supported by personalist leaders who would not respect the institutions of their respective countries (Dussel, 2007: 2–5). For Octavio Moreno, the destiny of populism, as a concept and theorisation, has become a chimera, "a fabulous monster, with the head of a lion, a goat body and a dragon tail", at the service of the forces and groups linked to neoliberalism. Any deviation from neoliberal orthodoxy is marked as populist. Any attempt to modify reality, the disaster that neoliberalism has led to in the region, is pointed out as populist. Thus, a concept with explanatory potential, linked to a particular historical experience, is voided and converted, the author tells us, into a political weapon, a "great other", a chimera, linked to demagogy, popular leadership and irresponsibility.[2] Kaiser and Alvarez (2016) highlight populism, in terms of the deception it generates, its state idolatry, which they favour over the market and individual initiative. Also, for its tendency to blame others for one's own ills and disabilities; the possession of an anti-neoliberal paranoia by means of which all existing ills are attributed to it; its democratic tendency, because it wants to discuss all republican institutions; and its egalitarian pretension.

Laclau takes a different view, perceiving it as a way to build political action and identity in a logical phenomenon with no clear boundaries. For Merlin (2014: 3), "unlike other authors who dealt with populism, Laclau does not start from the concept of the people as a given ontological assumption, but rather as a contingent effect, a particular political construction that has as a unit of analysis and originates in the social demand". According to Ranzani (2018), Laclau builds his perception of populism based on a discursive category as an effect of political significance. This phenomenon can develop in an authoritarian direction, but also into a democratic regime. It is a moment that arises around "equivalences" that lead to "popular hegemony". The dependence generated by populism is not a constant historical construction over time, although centred on consumer demand it does have a more stable expression. Derived from Saussure's theory of language, it supposes a construction of identity based on an articulation of demands. Following Freud and Lacan, it undermines the "mass" – barbarous, violent, impulsive and lacking in limits, hypnotised, with low intellectual performance, who seek to submit themselves to the authority of the powerful leader who dominates them by suggestion; a lot of people who do not constitute a social bond – while exalting the "people", who do not conform ideologically, morally or as a class, but from a logic that is generated articulating their demands. This is what, according to Laclau, gives him his identity as such, which relates and forms identity and, discursively, populism as meaning, based on the achievement of common demands and not on the charismatic obedience of a leader. Thus, populism would convert subjects into political actors, the opposite of what happens to the mass (Laclau, 2008). For Echeverría (2018), "The populist ideal would reside in an ideal situation of permanent referendums", and "the protagonist role claimed by the individual one by one, cheered by networks and media, has magnified the discomfort of individuals who feel excluded from the system, rebel against the 'status quo' and accept to play a gregarious rupturistic role in favour of a caudillo". He points to the pandemic of "democratic pathology", where populism always surrounds democracy and – if it can – supplants it. This "alter democracy" would be built on the rubble of traditional democracy. Cortés and Pelfini (2017) argue that it is necessary to analyse the attitudes of the intellectuals about populism. They ask if populism is the political "curse" of the continent, or if political studies about Latin America tend to reduce the political phenomena that do not fit in the Western imaginary of modern politics with populist logics. It is believed here that populism, being a political discourse, whose possibility of appropriation is wide, projecting itself in various ways, must be studied contextually. Therefore, this work will focus on Chile.

Populism in Chile

Caudillismo, *aristocracy, paternalism*

Norbert Lechner (2002) states that in Latin America order is not considered as a political problem, as a collective and conflictive work because "young republics rely more on the idea of the national state (and, therefore, a notion of community as a unity preconstituted) that in democratic procedures". Possibly because of this, populism has been transformed "into a central political category of public disputes" (Cortés and Pelfini, 2017: 58) in Chile. The elements that structure it, and that impede the development of enlightened democracy or other autonomous forms of government in the population in the twentieth century, are paternalism, which acquires, among others, a long-standing aristocratic and *caudillista* sense. For Medina Echavarría (1971, pp. 104–109), already in colonial times the uses of the paternalistic structure crystallized in the beliefs in a cordial value of personal relationships, in a shelter that could not be missing in a moment of crisis and in the unknown power, and for that reason unlimited, from the boss. Operating alongside it is a prevailing aristocratic sense. At the beginning of the twentieth century, its foundation lay in a series of religious beliefs and values that crystallized in a mythical vision of the world, which may even be aberrant or false from the point of view of the official Catholic doctrine. The most essential element of it is the split of humanity into two parts: the full, endowed with a transcendent and anointed conscience of the dignity of being made in the likeness of God, and the one that is in the making, tied strongly to the instinct and that requires redemption. Here is the natural superiority of some and the extrinsic inferiority of others (Vergara-Barros, 1978, pp. 230–240). The *caudillo* can carry these elements. The nineteenth century exercised a personal power sustained in dependent groups or clienteles that join him through a symbolic link, which represents the exchange of favours, protection, loyalty and support. It will seek to impose itself on local caciques covering the national scope. It bases its legitimacy in religious terms, since it is seen as having a magical power, and in the manifestation of special abilities, such as the use of weapons and war, which are the forms of expression of its dominion over things. Awakening these powers elicits fear and admiration at the same time. His clientele is linked to him in personal terms, not through the word or rational discourse, but through the rite that acts in function of a symbol and its representation (Ramos, 1985, pp. 45–55). These features survive strongly in the twentieth century, preventing, for example, the development of a deeper welfare state, because this refers to non-individual notions, collective criteria of democratic coexistence and economic discipline (Meller, 1998, pp. 311–318). Populism acquires these characteristics, in varying degrees.

From the Popular Front to Pinochet

The developments and interpretations of populism in Chile are varied. On the *ibañismo*, Jean Grugel (1992, pp. 169–186) expresses that it is democratising while Joaquín Fernández perceives it as nationalist, trying to unite the homeland to the people – common people, with tradition and virtue – around a leader, to generate an opposition to communism and oligarchy. For Fernández, these characteristics constitute it as anti-liberal and authoritarian. For Drake, Popular Front populism "channeled the wishes and mass campaigns of social welfare in support of an induced industrialization", made possible by communist conservatism and socialist populism, which opt for collaboration to avoid the chronic repression of previous regimes integrating itself into the political system and strengthening union organisations. "By admitting mass leaders, but not their ambitious programs, the dominant groups hoped to soften – not accentuate – the social-political conflict." Thus, "Chile developed politically allowing the democratic expression of social antagonisms", although their original causes were not eliminated because, although communists, and especially socialists, "may have endowed *mass politics with* the potential that would constitute it in a stronger agent of change – although gradual and limited – that process of political change had not yet proved that it could originate substantial social or economic reforms". In the late sixties, "Despite cultural continuity, paternalism seemed to be losing its power". The Popular Unity "broke *the* implicit *rules of the game* to a greater extent than its predecessors – the Christian Democrats – promoting profound reforms in order to respond to the social pluralism fostered by years of modernization and populist policies" (Drake, 1982, pp. 239, 240, 308, 309). Carlos Cousiño (2001, pp. 189–202) perceives the existence of a populist character of the Allende government in the economic aspect, where public spending increases due to political and ideological causes, generating inflation and scarcity of products. For him, it lacked an adequate leader to lead the process. It is important to mention, in addition, that the way in which the right perceives Jorge Alessandri and the Christian Democracy to Eduardo Frei Montalva is also inscribed in the *caudillista* tendencies mentioned.

The civic-military regime also shows populist characters. Alan Angel expresses it by questioning whether Pinochet's economic policy is a strict neoliberal model. He affirms: "Many of the privatizations that were made ... especially in its early years, transferred state monopolies to private mono-polies without an adequate regulatory framework, which constitutes a model of clientelist capitalism instead of being a neoliberal model." He adds that the government "showed a lack of interest in fiscal discipline, allowing populist expenditures between 1988 and 1989 to influence the outcome of the ple-biscite and subsequent presidential elections, causing inflation of 30%" (2014, pp. 13–14). It should also be considered that Pinochet was perceived as a *caudillo*. In the socio-political crisis of 1955–1973, the "crisis of inclusion" that Halpern poses to us, experienced in Chile towards the end of the 1960s (Friedmann-Lackington, 1971, pp. 426–427, 437–438), what was the state of Chilean society towards 1973 after more than a decade of uncertainty before the various political, social and economic experiments that were developed, especially in the face of a socially new event such as the systematic open vio-lence coming from, example, the press and street protests? And, along with the aforementioned, was not there the yearning for the establishment of an "order" that made everyday life more predictable? Were there elements that predisposed at least a segment of the society of that time to support an authoritarian order without too many traumas? Pinochet assumes elements of elitist group cohesion different because the democratic conditions that framed the exercise of the previous ones, the "President of the Republic" and the "Commander in Chief of the Army", are no longer the same. The fusion of both instances is becoming inevitable as of 1974, once the psychic legitima-tions and the mechanisms that allow a more optimal control of the country and, above all, of the Army, mature (Timmermann, 2005, pp. 165–330). Even though that year the *Declaración de Principios* maintains that there is a legitimation centred on the historical past of the country, the government of

Diego Portales and that therefore the government will "depersonalize" to "avoid" "all *caudillismo* alien to our idiosyncrasies" (Junta de Gobierno, 1974, p. 23). *Visión Futura* in 1979 refers to elements that enhance a particular leadership, the "human incarnation of the Portalian regime", "the institution of the President of the Republic", "authority that enjoyed very broad powers". It mentions the figure of Portales, "an implacable realist: not inspired by theories or books, but in Chilean society of his time, as she was, pushing the value of ideologies" (Pinochet, 1979, p. 2, p. 5). Pinochet did not exercise a special personal attraction in the masses,[3] but it does show "a remarkable power of persuasion" to attract leaders of the right, being able to involve in its project people belonging to diverse parties and movements (Huneeus, 2000, p. 132, p. 154). It is for Sergio Rillón a sort of divine envoy destined to end communism. It is possible that in the time under study, reiterating the words of Ramos (1985: 45–55), there continues to prevail in no small measure in some sectors "the vision of an ontologicalized social order, the vision of a hierarchical and class order, of sacred foundation".

Democracy and populism

Thinking of the second third of the twentieth century in Chile, populism emerges as a kind of transaction that is not agreed, and fully variable and contextual, in which political leaders and groups, clientelistic in their socio-political and economic projection, carry in their management authoritarian elements, paternalistic, elitist, oligarchic and *caudillista*, which nevertheless contradict their inclusive discourse in this sense, before groups that slowly populate the city. This is deepened when the ISI (Industrialización Sustitutiva de Importaciones) model begins to paralyse from the late 1950s. That is why populist practices emerge that ignore Enlightenment democracy, trying, instead, to emotionally insert the possible citizen into dependencies that project religious, political and economic insecurities.

They have referred to populist practices in which existing power flows from the top down, either from charismatic leaders or elites, without generating political autonomy that leads to full citizenship, constituting various ideological communities at times. That is why populism is perceived as a government technique in which, outside of liberal democracy, it seeks to manipulate the feeling of enlightened sovereignty in the population, organised or not, to achieve socio-political stabilisation objectives, incurring the risk, if necessary, to generate economic imbalances to meet their needs and demands. Possibly what you want to achieve is not precisely an autonomous mobilisation but to avoid it, contain it or give it a certain direction. Its ideological, linguistic, charismatic, etc. contents, are very diverse, and a democratic instance can be generated alien to the elites or charismatic leader, as Laclau maintains, but also may not occur – which is what has generally happened.

The desire to achieve the satisfaction of needs, a momentary identification with the charismatic leader, may or may not occur, which in no way involves expressing the talk of the exercise of enlightened sovereignty but rather of an economic or political-social calculation of survival. If a populist identity arises, as Laclau puts it in the demand, it is momentary and is a function of this satisfaction, which can also be perceived ideologically. It is believed here that it depends on the articulation of language with the level of survival need – which includes emotional accumulation – when one is about to cross the threshold of resistance to power. That is to say, it depends on how the emotional weight of the word activates the fear – the extreme insecurity that is suffered – to drive the action. This populist action does not necessarily have a rupturistic dimension of the system, posing a specific disagreement. Reigns also, as Laclau refers, the need for recognition by the other of something that has no institutional response. In this sense, thinking of further developments towards enlightened democracy, populism is a symptom of its weakness or non-existence, that its values have not been institutionalised nor taken to a particular introspection. Seen in this way, its existence is a danger to move further in this direction, but at the same time it is the verification that, in the socio-political stage of the second third of the twentieth century, it was possible to achieve this. This is the tension of the democratic "empty signifier" articulated by a group – by itself or being induced to do so – and that can lead to populism, but also to other political forms, such as fascism.

Neoliberal populism

Neoliberal terror

The existing horror after 1990–95 (Timmermann, 2016)[4] is deeper than that of the civic-military regime (Timmermann, 2015), in the sense of pleasure associated with pain. The rhythms of daily action are accelerated and compressed because one works more to consume more. Fun operates as an escape from such tension, not as a recreation of one's own. The spending on it of increasing sums of money legitimises the self-perception of being socially successful before others, of not being a "nobody" (Gambarotta, 2011). The individual escapes from the past by always throwing himself into the immediate future, in a constant personal flight that is not contained by the environment and that he does not perceive as such, consolidating an unstable identity. Socio-politically, any tension, problem or crisis is resolved from above, without enlightened sovereignty, or remains at levels sufficient to generate a *phobos* – uncontrolled and irrational fear – manageable *risk* that maintains stable levels of social instability and, therefore, enables one to maintain control.

Before the political and economic successes, already at the end of 1992, the new order was also legitimised by the remaining population, in a country that already in those years passed a per capita income of $3,000 and that began to see itself successful – Chile is the "jaguar" of Latin America. The actions required to be *happy* are work and consumption, as well as formally exercising a minimalist policy. The introduction of economic calculation guidelines as the basis of everyday behaviour and the production of meaning leads to a perception that the state is expected to satisfy economic and social goals as the core of what democracy is, and not the achievement of values linked to sovereignty. This would explain why, in surveys, people mostly express that they are *happy* in such a system (Pincheira, 2016, pp. 217–251). The existing insecurities are necessarily more related to expectations linked to unemployment, salary reductions, high inflation, stagnation of the economy, etc. and not to political aspects such as, for example, the reform of the 1980 Constitution and, today, the corruption of some sectors of politics, the Armed Forces and Carabineros, which in no small measure is consumed as a spectacle. As one does not participate politically to control these insecurities, it is evident that, in this respect, the existing discontent turns more inward than outward, reflecting a demobilisation and an intimate political disenchantment. This privatisation of fear and socio-political pain, this emptying of perceived social conflict, makes it difficult to perceive the existing *terror* and also its necessary conceptual rationalisation, as well as the socio-political aspects, as it is impossible to control their generating elements of uncertainty in terms of enlightened democracy.

Neoliberal populism

The Washington Consensus

Neoliberalism states that the economic and social intervention of the state must be reduced, allowing the market to regulate the relationship between consumers and entrepreneurs. All growth is guaranteed with balanced public accounts, not incurring expenses motivated by political causes as occurs, for example, in electoral clientelism. Public accounts should be balanced to avoid debt and uncontrolled inflation. The guidelines of these economic practices would be granted by the Washington Consensus from the beginning of the 1990s, but Felipe Morandé states that "The first two decades of this century have been marked by many contrasts in Latin America". He says: "there was a difference from the Washington Consensus in the populist governments of Argentina and Venezuela. Others like Bolivia (Evo Morales) and Ecuador (Rafael Correa), have maintained the essence of a market economy with good results, at least until before the fall in the prices of *commodities* in 2014. Peru, Colombia and Mexico have deepened in this direction" with a very positive economic performance. In Chile, he affirms that these reforms began in the seventies and that possibly inspired the OFM and the Washington Consensus. It was the first time that pro-market reforms were carried out in Latin America (Morandé, 2016).

For Oscar Landerretche, "During the early 1990s, any politician who promoted new or radical ideas for development would have been labeled demagogue or populist", which is why "negotiators of the transition to democracy focused on a single principle: the political and economic stability 'then' had a deeply redistributive character, due to the consequences that the uncertainty and economic volatility have among the poorest households". Siavelis argues that, despite the fact that the Concertación increased fiscal spending on social policies,

> they did not touch the private health and pensions system, nor have they made any attempt to find a redistribution mechanism that would level the unequal distribution structure of income and benefits opportunities to have a better education. They have kept the State as far as possible from economic activity, excluding it from the debate and implementation of any development strategy.

He adds that the Concertación must face up today to accusations of "having created irresponsible policies, fallen into populism and having had very little will to commit to a serious transformation of the economy" (Sehnbruch and Siavelis, 2014, p. 51, pp. 172–173).

If one thinks that populism has been characterised by spending more than the economic equilibrium of the country supports, and that this could have clientelistic or redistributive welfare-oriented motivations, it is difficult to be exhaustive in the classification of the Concertación. In terms of the fact that it strictly followed the parameters of the Washington Consensus, Landerretche maintains that this was not the case, since, on the contrary, "a very conservative banking regulation" was implemented by the Central Bank; structural and counter-cyclical budgets were established, promoting with them the fiscal deficit in periods of recession, without a structural deficit; and the welfare state was "slowly but surely" rebuilt. Subsequently, the exhaustion of the "stability imperative", he adds, "coincides with the adoption of this same political model by the conservative sectors. The political story that Sebastián Piñera delivered in his presidential campaign promised this same type of equilibrium and his failure was, evidently, the failure to provide them" (quoted in Sehnbruch and Siavelis, 2014, pp. 172–176). This means that the right discourse becomes populist and must govern through neoliberal programmes, different from its electoral campaign promises. Fundamental in this sense is to analyse the *"alcaldización"* of the policy because, already in the nineties, it is the base of the strengthening of clientelistic practices at the local level and of the neutralisation of the citizen's autonomy to accentuate dependencies. In this context, together with the emotional context of terror-happiness described above, it is in this that neoliberal populism must be situated.

Clientelism, local power and dependency

Neoliberal populism operates on the aforementioned emotional conditions, as well as others of a socio-political context. Marcelo Arnold mentions among these the discrediting of the institutional political system; the existence of a system of electoral representation coming from the civic-military regime that prevents the transparent generation of citizen sovereignty; the existence of a policy that becomes technical; the existent social inequalities, sustained privileges due to origin and not to merit; and the asymmetry of power between capital and labour, by weakening unions (Arnold, 2011). More specifically, Emanuelle Barozet perceives the existence of a political style based on a personalist leadership, sustained by clientelist networks that develop in local environments, where she perceives the existence of *caudillos* that establish non-transparent nexuses that allow them to distribute the public resources of their mayorships (Barozet, 2003, pp. 39–54; 2008, pp. 45–60).

In the nineties, the cases of Joaquín Lavín and Jaime Ravinet develop the aforementioned practices. In this regard, Veronica Valdivia argues that the reform of the early 1980s remained unchanged because the Concertación "confirmed the municipality as the entity in charge of materializing the subsidiary state and focusing on combating poverty as the redefinition of the policy". The management of Ravinet (1990–2000) favoured "the maturation of this process, deploying a technocratic leadership of neoliberal inspiration, which promoted among the inhabitants of the commune tendencies to depoliticization and de-citizenization", by means of "the consolidation of the subsidiary nature of the State, of the private sector as the engine of economic and social development and the local vision, alien to the great national debates, of the citizenship 'what' established important issues in the municipality, taking them out of the central apparatus, and enhanced personalization of politics in the figure of the mayors". The Municipal Council and the CESCO of the commune of Santiago "*participated* in the plans and communal decisions, but in a consultative and non-decisive way", concentrating the councils with the community organisations on issues and particular problems of each neighbourhood, favouring the concentration of the population in their own problems (Valdivia, 2018: 137).

Lavin, when he was mayor of the commune of las condes (1992–1999), "raised the supposed end of ideological divisions, the existence of a consensus in the social base of the neoliberal model", holding that to people, "what he was really interested in not deciding on alternative political projects, but on his daily and contingent problems ... efficient management, less bureaucracy and his personal well-being" (Álvarez, 2016: 41–61). Aníbal Pérez, who studied the case of the Christian Democrat mayor of Valparaíso (1990–2004) of the Concertación, Hernán Pinto, affirms that he used the policies focused on poverty, "through a solid apparatus" as mechanisms of clientelism, with the existence of "certain notion of pressure *from below* to achieve urban sanitation in the Buenos Aires hills" (Pérez, 2013: 89–113). In the case of Virginia Reginato, mayor of Viña del Mar, of the UDI, a right-wing party, she expresses "that it was a pragmatic attitude, but not a militant commitment to the popular world". It affirms that "the massive vote towards the mayoress has to do more with a personalist character, obtaining the gremialismo support to the figure more than to the party" (Pérez, 2014: 1–10).

These trends are maintained, with different emphases, already in the twenty-first century. The municipal administrations of the UDI mayors,

Carolina Plaza (2000–2011) and Vicky Barahona (2000–2006) in the municipalities of Huechuraba and Renca, studied by Álvarez, were based on clientelist networks, "capable of influencing the daily life of the inhabitants" "and in the personalization of politics", "vital to achieve breaking old political allegiances", in areas without a predominance of the right. Its operators were local "brokers" usually presidents of neighbourhood boards or leaders of other organisations. For this author, "the objective of depoliticization implied by this political style, was able to express itself in the pragmatism of the electorate, capable of voting for a right-wing candidate at the local level and another center-left candidate in the parliamentary or presidential ones" which meant "The validity of the old model at the national level, based on a logic closer to political-ideological considerations". In this way, "Political adhesion, at the beginning of the new century, became volatile and pragmatic" because "The financing model based on projects with limited time spans prevented the social organizations from being projected over time and strengthened the power of state agencies". These links "were strengthened by materializing in networks of trust between the community and the mayors. Field visits, telephone calls, concerns about the illnesses of people belonging to the Third Age or minors, the perception that what was promised was fulfilled, was vital for the transfer of ideological barriers in two communes of the same tradition left" (Álvarez, 2016: 41–61).

It is possible that, in the emotional construction of insecurity, and with it of fear, the previous dependencies will be consolidated if we consider what José Poblete calls the "punitive populism", a "vision of punishment" in "the States that adopted policies neoliberal economic 'with' a new approach to crime, which is characterized by a consensus in the political class around the control of crime", "a predilection for incapacitation as the desired effect of the penalty", seeking "the hardening of criminal and the retreat of the guarantees to satisfy the punitive hunger of the population" (Poblete, 2017: 1–39). The issue of delinquency, in the pre-electoral period for the presidential elections, was systematically projected to produce social fear in Chilevisión's newscasts, owned by Sebastián Piñera.

Neoliberal populism

Fassin proposes the thesis that populism is an instrument of neoliberalism, even when this populism calls itself "the left". It is a statement that acquires credibility to the extent that it is argued that neoliberalism provides a broad context for this. This point of emotional consensus is reached by the predominant development of terror and happiness, based on the accumulated effects of previous contexts. It is possible to understand the effect of the populist double illusion that neoliberalism imposes, as proposed by Fassin (2018): confusing the people with the popular classes to replace the ideological alternative between the right and the left, at the base of democracy, by a sociological and false opposition between the caste and "the people" and reduce the people to a mythical category and only knowable when it is their turn to vote in the elections, which erodes their representativeness. He adds that the "populist moment" is too similar to the conversion of people into consumers. This is understood if one considers that the daily economic tactics in the neoliberal context, the individual consumption for immediate personal satisfaction, is transferred to political practices, preventing the projection of any valuation of democracy, as already mentioned, exacerbating consumption and individual violence. In this act, in the bodily dependence of satisfying the appetite, the human being learns to ignore the socio-political pain of the Other, because this consumerist practice is absent from any projection of community social construction. He hardens himself because he hardens himself from others and remains a prisoner of the extreme management he can make of his body and time (Timmermann, 2018), in a production of subjectivity that, however, is introspectively extrospective. This leaves any definition of people with enlightened sovereignty meaningless. What neoliberal populism allows is a functional people based upon the acquisition of individual economic achievements in a society that is marked by inequality. That is to say, a populism that installs, with the body-individual, short-term protentions dependent on an occasional leadership, attending emotional communities of the consumer market that are variable in the material appetites that must be satisfied, always in a limited way. The short-term conjunctural placism of the satisfaction of demands by neoliberal populism is accentuated by the emotional implementation that it realises in the current technological conditions of propaganda and communications, accentuated with the immediacy of social networks and the use of cell phones, which it is possible to achieve due to the lack of massive political cultures granting community identity. Momentary agreements are produced that focus on certain needs that the institutions must satisfy. It is a consumer identity that is acquired. What predominated in the populism that developed in the twentieth century, before 1973, are dependencies linked to religious elements related to the aristocratic sense and *caudillismo*, some ideologically conceived, at a time when the transition from oligarchic authoritarianism to forms of enlightened democracy is made, in the midst of the beginning of an industrial modernisation and growing socio-political inclusion, which will later be ideologically projected for the realisation of "global planning". There, the degree of happiness possible in community areas is located and this does not ignore the Marxist or communitarian ideologies. Of course, populism tries to avoid the suffering caused by the uncertainty of the present-future through symbolic models linked also to practical actions correlated with the current social order as the original foundation. This generates an emotional dependence in the long term based on utopian desires of inclusive socio-political projects. The bases of dependence on populism before are not those of now, but their emotional impact as soon as they position themselves as dispensers of insecurities that lead to fear are the same. Today, these are those of terror, which hides other insecurities, those of the body. According to neoliberalism, postmodernity is the overcoming of modernity, so it is surprising to establish that a regime like the Chilean one, developed since the end of the twentieth century, has populist practices and tendencies. It means that the supposed socio-political neutrality of the market has failed to grant the expected social equilibrium and it would establish a new order of social inequalities, which "produces a rupture of belonging, as it implies the loss of positioning in a category to which it belonged, towards another uncertain place in the social structure", "the social exclusion of vast social sectors" (Hounie and Fernández, 2017: 102). As it is necessary for neoliberalism to prevent these groups from moving to emotional positions supported by anger and fear, from breaking their thresholds that lead them massively to actions of rebellion or electoral distancing, populist practices emerge for their control, such as those developed since the municipalities, supported by characters and clientelism. Part of this populism is appropriating the term, without naming it, since it rejects denominating populists, those who criticise the neoliberal minimal state, distance themselves from their postulates or propose other solutions foreign to the market. Neoliberal populism is discriminatory also because it discursively raises the extreme socioeconomic difference, by massively imposing structural unemployment, social limits on migrants – who are criminalised and diminished, blaming them for some social ills – and those who are not entrepreneurs, whom are perceived as lazy, incapable or useless and, therefore, deserving of being poor. It is the society of segments as a whole, divided, which, therefore, prevents the creation of a political community, except at times when that population, at the local level, is part of the clientele of a certain populist leader.

Without forgetting the contexts of occurrence and their urgent needs, first a populism developed in Chile that imposed from the beginning of the twentieth century dependencies located in the leadership of political elites of the left, centre or right linked to extreme poverty and its ideological interpretation. From the twenty-first century, it does so by the existence of individualisms centred on personal consumption that is realised in perceptions of terror and happiness, imposing dependencies from the body that are stimulated and manipulated by a neoliberal state that, from the municipalities, establishes clientelisms that prevent the generation of citizen autonomy on a larger scale. The populism of the twentieth century, as with the neoliberal populism of the twenty-first century, has prevented in Chile the end of that necessary transition towards a modernity that evidences more citizen autonomy and less dependence on any leadership to establish its socio-political objectives.

Notes

1 He argues that in the 1920s and 1930s populism, "although imbued with Marxism, emerged in Latin America as a coherent and widespread response to the initial acceleration of industrialization, social differentiation and the pressures of the middle and working class. The populists reflected and stoked those pressures by

promising simultaneous welfare measures and protected industrial growth" (Drake, 1992, p. 307).

2 Quoted by Aldo Fabián Hernández Solís (2017).

3 If charismatic authority – based on the belief in extraordinary gifts and attributes that some people have singular to the eyes of another group of people, a belief that carries a sense of commitment and unconditional follow – allows an individual to institutionalise a hegemonic system in terms of domination at the societal level, then Pinochet lacks "charisma", because it is based on other elements; I could point out that certain charismatic features are projected to establish a hegemony by consensus and not by coercion in their relationship with civilian and military elites.

4 In this paper the contextual elements of the socio-political, economic, cultural and emotional transformations of Chile after the civic-military regime are exposed.

References

Álvarez, R. (2016) 'Clientelismo y mediación política: Los casos de los municipios de Renca y Huechuraba en tiempos de la 'UDI Popular', *Revista de Historia y Ciencias Sociales Divergencia*, 5(6), pp. 41–63.

Angel, A. (2014) 'Prólogo', in K. Sehnbruch and P. Siavelis (eds) *El Balance. Política y políticas de la Concertación. 1990–2010*. Santiago: Editoria Catalonia Ltda.

Arnold, M. (2011) '¿Existen las bases para el populismo en Chile?', *Le Monde Diplomatique*, July. Online at: www.lemondediplomatique.cl/article1652.1652.html. [Accessed 28 April 2019]

Barozet, E. (2003) 'Movilización de Recursos y Redes Sociales en los neopopulismos: Hipótesis de Trabajo para el caso chileno', *Revista de Ciencia Política*, 23(1), pp. 39–54.

Barozet, E. (2008) 'Populismo regional y estado en Chile', *Estudios Interdisciplinarios de América Latina y el Caribe*, 19(2), pp. 45–60.

Cortés, A. and Pelfini, A. (2017) 'El populismo en Chile: ¿tan lejos o tan cerca?', *Izquierdas*, 32. Online at: http://dx.doi.org/10.4067/S0718-50492017000100058

Cousiño, C. (2001) 'Populismo y Radicalismo Durante El Gobierno de La Unidad Popular', *Estudios Públicos*, 82, pp. 189–202.

Di Tella, T. *et al.* (2008) *Diccionario de Ciencias Sociales y Políticas*. Buenos Aires: Editorial Emecé.

Dockendorf, A. and Kaiser, V. (2017) 'Populismo en América Latina: revisión de la literatura y la agenda', *Revista Austral de Ciencias Sociales*, 17. Online at: http://revistas.uach.cl/index.php/racs/article/view/997 [Accessed 28 April 2019]

Drake, P. (1992) *Socialismo y populismo: Chile 1936–1973*. Valparaíso: Ediciones Universitarias de Valparaíso.

Dussel, E. (2007) 'Cinco tesis sobre el populismo'. Online at: http://enriquedussel.com/txt/Populismo.5%20tesis.pdf. [Accessed 28 April 2019]

Echeverría, P. (2018) 'Populismo y antipopulismo; nacionalismo, neoliberalismo e imperialismo', in *La Haine. Proyecto de Desobediencia Informativa. Ciudad de México*. Online at: https://lahaine.org/aO3b [Accessed 28 April 2019]

Fassin, E. (2018) *Populismo de Izquierdas y Neoliberalismo*. Barcelona: Editorial Herder Editorial.

Friedmann, J. and Lackington, T. (1971) 'La Hiperurbanización y el desarrollo Nacional en Chile', in H. Godoy (ed.) *Estructura Social de Chile*. Santiago: Editorial Universitaria. Santiago.

Gambarotta, E. (2011) 'La dialéctica aporética del modo de corporalidad pugilístico: el control de lo natural y su descontrol', in V. Dhers and E. Galk (eds) *Estudios sociales sobre el cuerpo: prácticas, saberes, discursos en perspectiva*. Buenos Aires: Estudios Sociológicos Editora.

Giner, S., Lamo de Espinosa, E. and Torres, C. (1998) *Diccionario de Sociología*. Madrid: Editorial Alianza.

Grugel, J. (1992) 'Populism and the Political System in Chile: Ibañismo (1952–1958)'. *Bulletin of Latin American Research*, 11(2), pp. 169–186.

Hernández, A. (2017) 'El populismo, quimera para mantener el dominio neoliberal', *Revista de Ciencias Sociales Tla-melaua*, 41, pp. 247–249.

Hounie, A. and Fernández, A. (2017) 'Las desigualdades sociales y sus implicaciones con el sufrimiento contemporáneo', in *Políticas del dolor. La subjetividad comprometida. Un abordaje interdiscipolinario de la problemática del dolor*. Montevideo: Ediciones Universitarias. Universidad de la República.

Huneeus, C. (2000) *El Régimen de Pinochet*. Santiago: Editorial Sudamericana.

Junta de Gobierno (1974) *Declaración de Principios del Gobierno de Chile*. Santiago: Editora Nacional Gabriela Mistral.

Kaiser, A. and Álvarez, G. (2016) *El engaño populista, ¿Por qué se arruinan nuestros países y cómo rescatarlos?*Bilbao: Deusto.

Laclau, E. (2008) *La razón populista*. Buenos Aires: Fondo de Cultura Económica.

Lechner, N. (1987) *Los Patios Interiores de la Democracia*. Santiago: FLACSO.

Lechner, N. (2002) 'Nuestros Miedos', in J. Delumeau et al. *El miedo: Reflexiones sobre su dimensión social y cultural*. Medellín: Corporación Región.

Mackinnon, M. and Petrone, M. (2010) *Populismo y Neopopulismo en América Latina*. Buenos Aires: Eudeba.

Medina Echavarría, J. (1971) 'De la Hacienda a la empresa', in H. Godoy (ed.) *Estructura Social de Chile*. Santiago: Editorial Universitaria.

Meller, P. (1998) *Un siglo de economía política chilena (1890–1990)*. Santiago: Editorial Andrés Bello.

Merlin, N. (2014) 'Política y psicoanális: populismo y democracia', *Topía Magazine of Psychoanalysis, Society and Culture*. Online at: www.topia.com.ar/articulos/política-y-psicoanálisis-populismo-y-democracia

Morandé, F. (2016) 'A casi tres décadas del Consenso de Washington ¿Cuál es su legado en América Latina?', *Estudios Internacionales*, 48(185). Online at: http://dx.doi.org/10.5354/0719-3769.2016.44553

Pérez, A. (2013) 'Clientelismo Político, Neoliberalismo y La Concertación: el "guatón" Pinto En el Municipio de Valparaíso 1990–1996', *Revista Divergencia*, 3, pp. 89–113.

Pérez, A. (2014) '¿UDI Popular? Los Campamentos y El Respaldo Electoral-Popular de Derecha. El Caso de Virginia Reginato En Viña Del Mar (2008–2013)', *Izquierdas*, 21, pp. 1–30.

Pincheira, I. (2016) 'Las Encuestas de la Felicidad y la gestión gubernamental de las emociones en el Chile actual', in R. Rodríguez (ed.) *Evaluación, gestión y riesgo*. Santiago: Departamento de Psicología, Universidad Central de Chile.

Pinochet, A. (1979) *Visión Futura de Chile*. Santiago:División Nacional de Comunicación Social.

Poblete, J. (2017) *Populismo punitivo y neoliberalismo: una mirada crítica*. Memoria de licenciatura en ciencias jurídicas y sociales. Tesis Pregrado, Universidad de Chile. Online at: http://repositorio.uchile.cl/handle/2250/143956

Ramos, C. (1985) 'Caudillismo e Ilustración: Elementos Socioculturales para la Comprensión del Liderazgo Político en América Latina', *Estudios Sociales*, 22, pp. 45–55.

Ranzani, O. (2018) 'El populismo es un freno de mano al neoliberalismo'. Online at: www.pagina12.com.arhttps://www.pagina12.com.ar/91352-el-populismo-es-un-fr

Sehnbruch, K. and Siavelis, P. (2014) *El balance: Política y políticas de la Concertación: 1990–2010*. Santiago: Editorial Catalonia.

Timmermann, F. (2005) *El Factor Pinochet. Dispositivos de poder, legitimación y élites. Chile, 1973–1980*. Santiago: Ediciones Universidad Católica Silva Henríquez.

Timmermann, F. (2015) *El Gran Terror. Miedo, emoción y discurso. Chile, 1973–1980*. Santiago: Editorial Copygraph.

Timmermann, F. (2016) 'Great Terror and Neo-liberalism in Chile', in M. Korstanje (ed.) *Terrorism in a Global Village: How Terrorism Affected Our Daily Lives*. New York: Nova Publishers.

Timmermann, F. (2018) 'Implosion of Time: Body, Emotions, and Terror in the Neoliberal Civilization in Chile', in A. Scribano, F. Timmermann and M. Korstanje (eds) *Neoliberalism in Multi-Disciplinary Perspective*. New York: Palgrave Macmillan.

Valdivia, V. (2018) 'La "alcaldización de la política" en la post dictadura pinochetista: Las comunas de Santiago, Las Condes y Pudahuel', *Revista Izquierdas*, 38. Online at: http://dx.doi.org/10.4067/S0718-50492018000100113 [Accessed 28 April 2019]

Vergara, X. and Barros, L. (1978) 'La Imagen de la Mujer Aristocrática Hacia el 900', in P. Covarrubias (ed.) *Chile, Mujer y Sociedad*. Santiago: UNICEF.

9 The game of disillusion

Social movements and populism in Italy

Antimo L. Farro and Paola Rebughini

Populisms of the twenty-first century

Populism is pop today, but perhaps it is just 'old wine in new bottles'. Certainly, what surprises all the observers is its intensity and simultaneity in an increasing number of countries (Mudde and Rovira-Kaltwasser, 2017; Mudde and Rovira-Kaltwasser, 2019). In the last decade, 'populism' has become one of those overloaded, catch-all notions, whose stratified and intertwined meanings prevent any clear analytical use. Indeed, the poly-semantic nature of this notion is not new, but today there are even more cultural differences in its definitions, not only in respect to the traditional differences between continents, for example Latin America and Europe, but also in Europe and even within a single country. Moreover, the confusion is fostered by the presence of 'populism' in both ruling parties and minority political organisations.

The Italian philosopher Norberto Bobbio (1987) observed that populism lives within weakened forms of democracy, but never totally outside it. In the last twenty years, political analysts have described the metamorphosis of Western democracy in a form of 'counter-democracy', based on the mechanisms of negative politics (Rosanvallon, 2006), or in a form of 'post-democracy' (Crouch, 2000) in which democracy is in poor health and the hostage of lobbying practices. More recently, such poor health, well captured in the definition of 'democracy fatigue' (Appadurai, 2017), has led towards more explicit forms of populism, structured around the definition of 'popular will'.

Hence, today, the notion of populism cannot be apparently excluded from our vocabulary. It is useful for understanding some of the major cultural changes of the early twenty-first century, such as the crisis of trust in the processes of globalisation, the rise of communitarian identities against the possibility of multiculturalism, the mixing up and the partiality in defining who the 'people' deserving rights are. These social and cultural transformations have been accelerated by the 'great recession' started in the US in 2007 and the consequent acceleration of the process of social inequalities already present in all the Western world, fostering current forms of 'great regression' (Geiselberger, 2017). The observation that our social relations and social bonds are increasingly the product of technological networks and economic interdependences, rather than of choices based on values, nourishes a political offer capable of intercepting the nostalgia for communities and reassembling identities (Martuccelli, 2017).

The word 'populism' is usually brought into the media debate to highlight the fracture between some kind of 'élite', or 'oligarchies', and those not belonging to an alleged privilege stratum of society, 'anonymous masses', 'common people,' 'folk', who no longer think themselves as belonging to the working classes, while work has become an abstract, opaque notion in itself. Moreover, this idea of 'common people' does not include immigrants, or minorities, and looks for its roots in local identities. Today, the need for identity, recognition and protection of the people included in this non-elitist, amorphous, community seems satisfied by a political offer – organised by political parties and social movements – able to mobilise through the reassembling of passwords, scapegoats and virtual spaces of mutual recognition, in the absence of real social spaces (Caruso, 2018; Diamanti and Lazar, 2018; Urbinati, 2019). This is possible also thanks to current communication and digital media that foster a direct relation between the leader and the people, allowing the 'audience' to feel they have a voice. Social media boost populist communication styles through platforms suited to producing emotional and superficial comments, as well as reconstruction activities, including fake-news (Mazzoleni and Bracciale, 2018).

On the one hand, the traditional dichotomies of modernity have been wiped away. The desire for a direct participation – in the Web or in public space – is not contradictory to the trust in a charismatic leader; to be believers in the values of individualism and meritocracy can be combined with the claim for more social equality; the trust in direct democracy is not contradictory to the claim of a strong state; and the research about the new is not in contrast with the return of pre-modern reactionary traditions. On the other hand, this new form of heterogeneous and contradictory 'totality' – antithetic with the Hegelian modern totality made by synthesis – is based on a radical opposition to any sort of 'otherness', first of all the immigrant and any 'élite'. It is an opposition focused on the present, and unable to recognise the genealogy of social phenomena. Contemporary populism, well rooted in the mechanisms of digital media, lives on contradictions, flattery and immanence; it occurs in the absence of memory of what happened yesterday, and in the eschatological promise of a newness that is no longer emancipative, but just reassuring, again based on the promise that economic growth and social mobility will not stop. In this respect, populist culture is a 'totality' unable to imagine the future, because its game is to play with the disillusions of the present.

In the following sections, we present a cartography of the main characteristics of current forms of populism, and of its implicit historical convergences with the requests of contemporary emancipative social movements; then we introduce some of the main characteristics of the Italian case. Indeed, current Italian populism has been presented mainly as related to the Five Star Movement and League party, and to their political opportunism. In the chapter, we explain the

genealogy of Italian cultural characteristics, with a specific attention to the relation between individual political participation and communitarian temptations.

Social movements and the attempt to create new political spaces

The agencies of twenty-first century collective movements and populism in
the Western world have similar issues of systemic structuration. However, they
act in two opposite ways: they signify opposite meanings to these issues and
have opposite attitudes when transposing these issues into political action.
Collective movements such as 15-M Indignados and Occupy criticise domination by global systemic forces (such as global financial forces) and pursue
an alternative structuration of social life. However, they do not translate their
action into political forms. The populist forces, on the other hand, defend
either local or national interests against economic globalisation, criticise
establishments, defend Western values against multiculturalism, reject immigrants and, in the European Union, foster the disillusion towards the Union
and the eurozone. The issues at stake in the confrontation between these two
opposing socioeconomic, cultural and political orientations are of key
importance to understand the complexity behind our everyday life. Both are
expression of a post-industrial society, accustomed to neoliberal culture and
entrapped in a long-term crisis of modernity (Touraine, 2007), and their
opposed political tenets arise on the same ground of social inequality, ecological risk and technological change.

Agency and subjectivation inside social movements

The emancipative collective movements of the twenty-first century embody the
attempt to shed light on the specific forms of domination characterising our time
through some major issues: a) the search for alternatives to the systemic forces
like those of global finance, and their conditioning of the existence of individuals
and groups; b) the strong meaning attached to the uniqueness of the individual's
subjective involvement in collective action; c) the recourse to online communication and experiments in new modes of inter-subjective relationships as support for new forms of democracy, both for organising their initiatives, as well as
prefiguring a different political system focused on the dignity of everyone; d) the
construction of new spaces of social life in which it is possible to experiment with
new forms of sociality; e) the coping with the fragmentation perceived both by
activist and citizens in respect to the separation between systemic forces – such as
global finance – and everyday life.

In reference to these issues we can define a collective movement as a
common action conducted by subjects aiming to affirm themselves as self-
directed actors, pursuing universalistic alternatives to dominant cultural,
economic and social systemic orientations. These subjects constitute a
common action which seeks both to identify and challenge its opponents –
the systemic forces conditioning systemic development – and to control the

direction of these systemic orientations. The collective movement, then, constitutes a level of initiative which aims both to construct conflicts with systemic forces and to integrate these problematic relationships through the regeneration of institutional systems (Melucci, 1996; Farro, 2000; Touraine, 2007; Castells, 2012; Farro and Lustiger-Thaler, 2014).

In this respect, the collective movements challenge the hegemonic representations connected to the construction of social life, currently centred on neoliberal culture. Again, this kind of struggle is not new and rooted in the 1970s with new social movements, at a time of the progressive fall of industrial societies and their welfare state systems, fostering more attention to the individual. This happened in an international context of increasing globalisation and enhancing neoliberal policies on the wave of China's political overture to liberal economy, the Soviet Union's collapse and the end of Cold War (Findlay and O'Rourke, 2007). These events had well-known repercussions also in theoretical studies. The idea of the ineluctability of modernity was radically criticised by the postmodern perspective (Lyotard, 1979; Rosenau, 1992). The historicist confidence in solidarity was abandoned in favour of an 'ironic' individual attitude (Rorty, 1989). Deconstruction and negative critique became the main theoretical offer in face of the crisis of modern narratives (Rebughini, 2010). Indeed, other critiques of modernity were also approached by scholars engaged in the effort to rethink modernity – such as Habermas and Touraine – in the analysis of the new era as a form of reflexive modernity – such as in the case of Beck, Giddens and Lash – in rethinking the possibility of justice in contemporary world, such as in the case of Rawls and Sen.

Globalisation, neoliberal culture, cognitive capitalism and necessity to overcome a merely auto-confutative critique of modernity are at the forefront of the ambitions of the waves of mobilisations that were born from the 'Great Recession' of 2007 and from the impulse of the Arab Spring. The 15-M Indignados movement and Occupy are the two main protagonists of this moment, at least in the Western world. May 2011 saw the launch of the 15-M movement of the Indignados in the EU, and September of the same year, the founding of the Occupy Wall Street movement in the United States. The two initiatives involved a denouncement of neoliberal finance and other systemic forces of globalisation, accusing them of conditioning all the aspects of individual life, and compromising democracy by subtracting its activities from institutional control.

The 15-M Indignados began in Spain with the mobilisation of thousands of precariously employed people from a range of social classes, many with well-qualified skills. It denounced the financial powers and the crisis they were causing by creating unemployment, rendering individual life precarious – especially among the youth – and fragmenting social cohesion. The mobilisation took the form of public protests and online communication networks, through which the movement's organisational circuits were strengthened, without giving rise to an established leadership. Similar mobilisations occurred in other EU countries, such as France and Italy, but lasted only a few months, while in Spain they have been ongoing since the beginning of 2014

(Castells, 2012). Meanwhile, in the same year, the Occupy Wall Street movement started in the United States with the establishment of a protest camp in Zuccotti Park, in the heart of New York's financial district. The protesters came from a wide range of social and educational classes, most of them high. Their action had a strong media impact and quickly attracted a vast number of people, with protests held in a number of North American cities, as well as on social networks. The movement opposed systemic power symbolically represented by the financial techno-structure of Wall Street, denounced as a constituent part of the 1 per cent of the population whose speculative activities conditioned the lives of the other 99 per cent, generating precarious and unequal living conditions.

With their mobilisations, participants in the Occupy and 15-M Indignados movements construct a resistance that underpins an affirmation of the right of every individual to escape systemic powers and forms of interiorised domination in everyday life. In Touraine's words (2007), this takes the form of an assertion of individual feeling, understood as the universal right to dignity of the human being to reflect on his or her situation and affirm a desire to exercise direct control of the evolution of his or her existence. Hence, this involves the resistance of an individual who presents himself or herself as an Occupy or 15-M Indignados protester, engaged primarily with a process of emancipation. Subjective engagement in these collective initiatives therefore takes the form of a 'subjectivisation of collective action' for which the role of the subjective struggle is paramount for the construction of the collective action; the singular subject with his/her own uniqueness precedes the development of a collective identity and practice which results from the connections of such subjectivation processes (Melucci, 1996). Singular individuals construct understandings among themselves aimed at affirming everyone's right to free themselves from powers deriving from techno-structures.[1] However, there is no relationship, not even a conflictual one, between the latter and the individuals or groups who constitute movements such as Occupy and 15-M Indignados. In fact, the protesters or their representatives do not engage in any contact with the representatives of these forces in order to negotiate or explain their antagonistic positions within an institutional framework. Rather, between protesters and techno-structures there are technologies accessed only by those with the skills or resources to do so, as is the case with those who, inserted within the financial techno-structures, develop financial flows on a global scale through resorting to online technologies.

Consequently, no collective movement or individual is in the position to directly confront these techno-structures, which operate without having to deal with a true democratic debate, addressing themselves to an arena of economic interests or institutional interests such those of foreign defence or energy supply. Such debates are formally absent also at a national, European or meta-national level (such as the UN) where representatives of the techno-structures work as lobbies. In this way, the fragmentation of social life is connected to the absence of any institutional integration between systemic forces and individuals who suffer their dominion.

New spaces of social life

In their short life, Occupy and 15-M Indignados movements created spaces for experimenting with alternative ways of living. These spaces were created and delimited by collective mobilisations in which individuals experimented with ways of emancipation from systemic domination and conditioning. The extension of these spaces was also fluid, and limited in time. They were either physical spaces that correspond to the places where protests, assemblies or other meetings occur, or virtual communicative spaces composed of online communication networks. These spaces of social life were the arena where people could meet on the basis of common concerns, mutual recognition, a common interest in understanding the mechanisms of current power relations, and the desire to practice common activities based on the principle of valorisation of individual dignity and personal capacity. These spaces of encounter aimed to tackle the fragmentation of social life and to experiment with new forms to valorise personal subjectivation in a context of cooperative action. The spaces of socialisation opened by the mobilisations of Occupy and 15-M, and by other local minor initiatives had – and continue to have where they are still in place – the explicit aim to contrast with the neoliberal culture of individualism, self-management, competition and adaptation of the fittest (Fraser, 2017). At the same time, they represent a radical alternative in respect to the collectivist social space of the industrial society and left parties of the past, where little room was reserved for subjective uniqueness and creativity. In the current social spaces of life, the individual subject is valorised as the fundamental element of a collective practice of resistance; individualisation is fully part of the collective action, and the new forms of resistance as everyday practices could not exist without the recognition of subjective uniqueness (Beck, 2016).

On a political level, the space of social life created by Occupy and 15-M movements involve the construction of a living democracy, as an everyday practice of mobilisation. This was realised through a mix of cultural, social and political activities, organised on the basis of direct democracy in terms of shared decisions and on the basis of mutual recognition in their concrete realisation, hence avoiding hierarchical configurations and charismatic leadership. Again, the valorisation of singular subjectivities and the capacity for sharing individual resources for a collective aim represent the political culture of these new forms of mobilisation. And yet, there has been no translation of these forms of collective action into traditional forms of institutional politics. The gap between mobilisations in everyday life and institutional political decision has not been so wide since the end of Second World War. Indeed, the questions raised by Occupy – from social inequalities to environmental risks – have not been taken into account by the national political systems, nor have these questions been discussed in international arenas. Moreover, an attempt to transpose the 15-M movement into the political party Podemos merely created a new form of 'left-wing populism' (Mouffe, 2018).

The black box of complexity and the rejection of globalisation

Twenty-first-century Western populism consists of traditional or recent political forces that consider globalisation's contradictions as a good standpoint to promote their defence of local or national interest, fostering conservatory or frankly reactionary politics. While emancipative social movements of the last decade, such as Occupy and 15-M, were also developed on the basis of a critique of neoliberal globalisation, and its consequences in terms of inequality and environmental crisis, their form of left-populism – as Chantal Mouffe calls it (2018) – is obviously completely different from the right-wing one.

Western experiences of contemporary populism present political attitudes opposite to the political experiences of collective movements and they proceed in an inverse direction compared to those collective movements that unsuccessfully attempted to construct a new political arena and were unable to translate their proposal in institutional political action. As we will see, one of their main differences – beside their obvious differences in terms of values – concerns the role of the individual subject inside the political action.

Contrary to social movements focused on everyday life, shared spaces of recognition, local activities and valorisation of individual uniqueness, from the start populist forces compare themselves with traditional and institutional political life. Overall, they seek electoral consensus focusing on some precise focal point: a) to denounce the precarisation of the social condition of their potential voters; b) to criticise the 'elite' or the 'establishment' as opaque and polymorphous social strata considered as responsible for this precarisation; c) to ask for the overcoming of free trade treaties (including those inside EU) accused of producing economic uncertainty, social inequalities and loss of national identities; d) to criticise globalisation as a whole, to defend national and local cultures and production; e) to produce a clear separation between citizens and unwelcome guests (immigrants, even those already resident in the country), considering rights as a scarce resource to be distributed in a hierarchical way; and f) to create enemies and scapegoats whose removal will produce a new equilibration of social resources.

We can easily identify these positions in the US, in Europe and also in many non-Western countries such as in India or in Latin America. The naive celebration of globalisation, the auto-confutative critique produced by postmodern culture and the incautious appeasement of social-democratic parties towards neoliberal culture has led to a cultural and economic turning point exacerbated by the Great Recession of the last ten years and its long-term consequences. In the face of an indecipherable complexity of contemporary society – in terms of production processes, techno-science developments, financial interests, global politics and intercultural relations – and in the face of the uncertainty of the short-term future, populism offers simple answers, while emancipative mobilisations want to open up the black box of 'complexity'. This divergence is particularly evident when we look at the situation of the younger generations (Colombo and Rebughini, 2019).

The reassuring promises of populist rhetoric displace the attention from the complexity that characterises the current immaterial and knowledge economy, instability of work and wages, injunction to be always performative, lack of trade-union support, and new forms of personal and social vulnerability, mainly associated with the precariousness of employment (Castel, 2016). While reassuring through the proposal of a monistic identity and the identification of a polymorphous enemy, populist rhetoric is unable to criticise the biopolitical matrix of the neoliberal system or the distortions of the liberal culture of individual merits (Cingolani, 2014).

While populist discourse focuses on the immanence of target problems, whose political maximisation is fostered by current social networks and the rapidity of the flux of information, social mobilisations creating space of shared social life aspire to freeze such acceleration. Sharing activities in mutual and horizontal recognition is also a way to break the 'iron cage' of immanence and 'presentification' and take the time to analyse the situation. The transformations of production and consumption have clearly been accelerating in the last decade, fostering the erosion of subjective experiences, and leaving social actors in continuous feelings of anxiety (Rosa and Scheurman, 2009). Recalling an alleged golden time of the past, the populist rhetoric gives the illusion that expectations based on past experiences of 'tradition' will reliably match the future. In a productive and techno-scientific landscape that boosts the compression of temporal space, transforming it in a task-oriented time, nostalgic reconstructions of an alleged ordered past give the paradoxical impression of newness.

The success of this cultural operation – after twenty years of defeats of the claims of alter-global movements, from global justice movements to Occupy – is evident in its capability to reorganise the same criticisms into a conservative proposal assuring the survival of these very same economic mechanisms. While alter-global movements or Occupy proposed a complex criticism of neoliberal economy, and its social and environmental disequilibrium, without abandoning the cultural enrichment brought by globalisation, current populist proposals attack particular economic agreements or engage in commercial struggles in the name of national defence. Their promise of an alleged social equality is based on the defence, and the enclosure, of one's own community. While alter-global and Occupy based their mobilisation's culture on the personal capacities of the autonomous subject, on mutual recognition of individual dignity, conservative criticism of globalisation constructs a communitarian culture based on Carl Schmitt's alternative of friend/enemy (Schmitt, 2005 [1922]). In pointing out an enemy, populism seems to offer more cogency that emancipative movements, with their ambition to analyse the complex reasons for inequality.

In a very old fashion, in Europe this logic is extended to the boosting of nationalism. In spite of their different national origins, populist political proposals base their identity on the defence of national integrity in face of the upper classes of 'supra-national technocratic elites', and the underclass of immigrants. In the case of Europe, in each country these political forces have different stories, a different intensity of internal heterogeneity and different pathways of political convergences. Some have roots in the totalitarian cultures of fascism and Nazism and explicit racist claims (Copsey, 2008); others consider themselves as 'post-fascist parties', such as the French Front National; other proposals are new political entities raised from the heterogeneous critique of globalisation, such as Britain's United Kingdom Independent Party (UKIP), and in Italy the Five Star Movement (M5S). The Italian situation distinguishes itself in the current convergence of heterogeneous forces such as the M5S and the League composing complementary versions of a diversified populist culture.

The Italian case

In Italy, the story of the difficulties, and often the failure, of emancipative social movements in having a systemic and institutional impact in terms of reforms is long and complex (Melucci, 1996). In this short text, only a brief account of the last twenty years can be made. On the one hand, we can recall the success of the Italian alter-global movements in the first part of the 2000s (Farro, 2006; Farro and Rebughini, 2008), and on the contrary, the failure of the few attempts to import Occupy mobilisations in the following decade; on the other hand, we can notice the specificities of the Italian 'post-fascist constellation', with its capability to adopt old right-wing themes in an updated form, while maintaining some of its extremist language, but tailored to the confines of the democratic terrain.

In terms of the reformist attempts at mobilisation, the Italian case presents an anomalous situation in which we can register a lively moment of critical and creative alter-global mobilisations (Farro, 2006; Farro and Rebughini, 2008), around the first years of the 2000s, and then a tiny reactivity to the impulse of international mobilisations a decade after, with Occupy and the Arab Spring. More than by an explicit mobilisation against the culture of neoliberalism, the Italian case is characterised by a diffuse and capillary mobilisation based on everyday practices of resistance, in an attempt to link the individual experience of the activist to the collective experience whereby a form of solidarity and reciprocal recognition is elaborated (Melucci, 1996). In the phase of alter-global movements, as in the previous 'new social movements' of the 1980s, critique and resistance are expressed in a variety of forms of mobilisation that involve both the political and cultural spheres.

This culture and practice of mobilisation is a far cry from populism as it is not communitarian, but based on the subjectivity of the activist. The moment of individual critical capacity is when each actor expresses and mobilises his or her sense of injustice and consequent resistance to forms of domination in different social areas (ecology, work precariousness, gender issues and so on), associated with a need to share experiences. In the space of everyday life, through their habitual network of social relations, the individual starts to elaborate his or her sense of indignation and need for resistance, to gradually modify some of his or her daily practices and life choices. The conflicts that alter-global movements create do not regard the competition for the means and resources to enter into a decisional and political process, but have been more orientated to the production of new practices and symbolic goods, such as critical consumerism (Rebughini and Sassatelli, 2008). Sharing a space of self-reflexivity, of negotiation and critique, of experimentation with new practices is different from the politics of the enemy/friend fostered by populism. In this respect, the culture of everyday life, as a typical space of emancipative local action, characteristic of the legacy of Italian emancipative movements, might partially explain the absence of a direct passage from this legacy to forms of left-wing populism in this country. In Italy, it seems that the more the mobilisations are localised, capillary and focused on individual everyday practices, the more they are proofed against the communitarian logic of the 'us' and 'them' typical of populism. Indeed, everyday practices highlight the role of the individual subject, of local differences, of the needs for adaptation and translations, rather than monistic references to a common identity.

This can perhaps explain both the modest participation of Italian emancipative movements to the cycle of Occupy international mobilisations – with their specific focus on the economic crisis – and the lack of left-wing populism more visible in other European countries. Certainly, in Italy, this situation has left all the field of populist culture to the right-wing organisations.

On this side of the populist constellation, we can recall the transformations of the Northern League – now called simply the League – from a regional secessionist party, to a national populist party (Biorcio, 2015, 2010), and the specificities of the heterogeneous populism of the Five Star Movement (Biorcio and Natale, 2018). Founded in 1991 by regional groups active in northern Italy and Tuscany, inspired by the example of the Front National in France, the populism of the League saw the expansion of consensus from the northern regions to the whole country. Ironically, the narrative of this party started off with the stereotype that southern Italians are less civilised and hard-working than those from the north. Yet, even though this regionalist vein is still present, in the claims of more regional autonomy in respect to economic resources locally produced, this is not contradictory with a national narrative based on a strong unifying leadership. All this required a redefinition of the previous regional positions with the adoption of a language of neo-patriotism in defence of the community of Italian citizens against the political, cultural, social and economic interference of globalisation, European technocracy, global financial flows, fear of immigration.

In the new narrative, the economic crisis and its social consequences are presented as a product of the loss of national sovereignty, rather than as a product of economic choices whose responsibilities are largely distributed among different national and international institutions. On the side of the media, the exposure of Italy to immigration influxes from the south of the Mediterranean Sea, and the ubiquitous images of disembarkations on the everyday news, offer a great political asset to capitalise frustration and resentment. Hence, immigration is denounced as one of the most serious problems facing the country, described as the one that most directly affects the poorer classes, deteriorates urban neighbourhoods and increases competition in the labour market. Together with Euroscepticism and hostility in respect to the euro as a common currency, this has led to electoral successes, largely thanks to working-class voters.

What is the role of the individual subject in this particular case? Formally, this discourse in defence of the Italians does not focus on citizens as individuals, but as members of a community. The problems related to unemployment, social inequality or inefficiency of institutions, are not singularised; the virtual representation of a community of belonging helps to contrast the individualisation of systemic problems and to compensate the disappearance of class references (Beck, 2016). The single citizen can be reassured by the sensation of being part of a communitarian block united against the same enemies. For example, the individual who manifests hostility and refusal in front of immigrants constructs his/her subjectivity in negative terms, what s/he can be is based on what s/he do not want to be; his/her dignity is possible because others have no such dignity. By imagining a reality – an integral community – that allows a monolithic coherence, this affirmation of subjectivity simultaneously affirms the individual and his or her sense of rebellion against political, cultural, economic and social exclusion. This passage from the anxiety of singularisation (Martuccelli, 2017), to the reassuring common belonging to an imaginary community (Anderson, 1983), also operates with respect to the problems of social inequalities and unemployment. The reference to the local and national community blurs the internal differences and hides the contradiction of the denouncement of technocracy while maintaining a neoliberal economic policy.

The other component of current Italian populism is the Five Star Movement (M5S). This is a recently constituted Italian political party that has an important parliamentary group in the Republic's Chamber of Deputies and Senate, elected members in the European Parliament, elected councillors in various regions, as well as mayors of important cities such as Rome and Turin. In the national election of 2018 it got a significant share of the vote and become the biggest party in the country, and set up a government together with the League.

The M5S is largely recognised as a populist entity and it proposes the classical criticism against globalisation and technocracies, even though its political discourse is highly heterogeneous, and after a few months of national government appeared to be contradictory on many issues. The differences with other European populisms concern at least two characteristics. The first consists in the modality with which it addresses the key questions of immigration and the presence of Italy in the eurozone. M5S criticism of immigration is more ambiguous and it changes according to the specific situations and the media debates. Immigration is usually presented as a problem for the job market in terms of competition, rather than as a cultural danger.

Regarding the presence of the country in the eurozone, the M5S manifests a wavering scepticism, and it promoted the idea of a referendum in order to decide the issue. The second difference concerns the accent on the denunciation of corruption and privileges of political and technocratic élites. Much of the electoral success of the M5S was related to its capacity to connect its political proposal with at least two decades of Italian debate about the corruption of the ruling classes.

Moreover, the most relevant originality of M5S in respect to other forms of European populisms is related to its organisational model which, inspired by the collective movements of the last two decades, presents itself as a horizontal articulation of an online communicative process where, in principle, every participant has the same value as any other member of the political formation. At the same time, there was a strong symbolic leadership of a charismatic founder – the comedian Beppe Grillo – and an important role of the managers of the informatics platform on which activists can vote online. Overall, this fostered a rhetoric for which the single subjects count for the implementation of M5S aims and values with the help of tools of 'direct democracy'.

This means that – contrary to other forms of populism – M5S presents a political and cultural programme for which the individual and not the community counts. The formal funding project of the movement was to boost individual political participation through technological tools such as the informatics platform for voting. Yet, this idea of direct democracy was never associated straightforwardly with anti-austerity movements, rather with a rejection of the existing political structures, accused of inefficiency and corruption. Hence, the individual's participation was more related to a protest against 'the elite' than to an emancipative proposal of social equality. The individual is able to express his/her frustration, feeling part of a horizontal network of other voters, while the political project remains blurred.

This can explain why the proposal for individual participation was born on the local scale, where local leadership could reassemble activists under some more precise aims. Among these aims – at least at the beginning of M5S adventure – there were also ecologist and emancipative proposals. Nevertheless, the desire for honesty and cleanness of the political class remained prevalent, and associated with a more conservative mood, with the reference to a sort of mythical past when things worked properly. This has fostered the typical Italian distrust towards public services that do not work efficiently. Consequently, individual participation tends to turn to individualisation in the sense of a neoliberal culture, rather than as a collaborative participation in sharing spaces.

Conclusions

The emancipative collective movements of the last decades and current forms of populism are opposing phenomena sharing common roots. They are both related to the de-structuration of preceding societies, and to the challenge to construct new global frameworks of social life and political systems. Focusing on the Italian case, we have traced the parallel pathways of emancipative social movements and populist politics as possible responses to the need and questions raised by the end of the trust in globalisation processes, and by the increasing inequalities brought by neoliberalist politics and the consequences of economic crisis. In this analysis, we have concentrated our attention on the role of the single subject and his/her emancipative autonomous action in this scenario. What seems to emerge is a clear alternative between the possibility of sharing participative activities, based on autonomous subjects mutually recognising their dignity as persons deserving rights, and communitarian choices in which a subject in crisis searches for recognition and protection in an imagined community.

According to those who, like Mouffe (2018), supported the necessity of left-wing populism (a populist strategy for which it is possible to clarify the 'us' and 'them' of the political arena), populisms of all kinds are inevitable consequences of thirty years of neoliberalism. The latter led to a process of construction of oligarchies, and to the exacerbation of social inequalities involving a large stratum of the population, from middle to lower social classes. Active political participation was also belittled, leaving citizens feeling themselves without voice, and a strong scepticism towards traditional parties flooded, in a situation of 'post-democracy' (Crouch, 2000).

According to this interpretation, the sharing activities of single citizens or groups working in the space of everyday life is not sufficient, since the struggle between 'us' and 'them' requires clear collective identities. This can perhaps explain the initial success of M5S in Italy, and its capability to intercept the need for participation of the atomised citizen. In Italy, after a decade of weak visibility of the action and voices of emancipative social movements, the populist political offer of M5S and the League has reintroduced the conflict in the political arena – in response to the previous neoliberal adagio of the widespread belonging to the middle class adopted, among others, by the Democratic Party. In Italy, the legacy of social movements of the last twenty years has evidently not produced a left-wing populist front as in other European countries (for example, in Spain or France), and the scenario appears as much more fragmented. Hence, at least at the moment, in Italy there is not the conflict between populist oppositions that we can see in other European countries.

To sum up, we believe that from a sociological perspective it is interesting to observe the role of individual subjects in the political dynamics of populism, not only in respect to their choices of belonging, but starting from their chances to express themselves as autonomous subjects, with personal capacities, and capabilities of mutual recognition. This quick analysis of the Italian case can show that the success of populist politics is not only related to a systemic and structural crisis, but also to what happens at a subjective level, to individualisation processes, and to the way in which an individual can express his/her autonomy in the public and private space.

Note

1 The concept of technostructure is adopted here by critically taking the term proposed by J. K. Galbraith following its analytical perspective to define the leading actors of the new industrial state (1967). It refers to the organizational structures of agents mobilizing the technological, economic and scientific resources concerning the power development of the systemic forces, as in the case of finances' systemic forces. The activity of the technostructures – as in the case of the financial flows – conditions individual and group life, while this activity separately develops from the actions of these individuals and groups. This can be verified by means of technological platforms or management infrastructures for scientific knowledge which are interposed between the systemic forces technostructures' activity and the individuals and groups living experiences (Touraine, 2007; Castells, 1996).

References

Anderson, B. (1983) *Imagined Communities: Reflections on the Origin and Spread of Nationalism*. London: Verso.

Appadurai, A. (2017) 'Democracy fatigue', in H. Geiselberger (ed.) *The Great Regression*. Cambridge: Polity Press, pp. 1–12.

Beck, U. (2016) *The Metamorphosis of the World*. Cambridge: Polity Press.

Biorcio, R. (2015) *Il populismo nella politica italiana: Da Bossi a Berlusconi, da Grillo a Renzi*. Milano: Mimesis.

Biorcio, R. (2010) *La rivincita del Nord: La Lega dalla contestazione al governo*. Roma: Laterza.

Biorcio, R. and Natale, P. (2018) *Il Movimento 5 stelle: dalla protesta al governo*. Milano: Mimesis.

Bobbio, N. (1987) *The Future of Democracy: A Defense of the Rules of the Game*. Cambridge: Polity Press.

Caruso, L. (2018) *Populismo contemporaneo*. Fondazione Feltrinelli papers.

Castel, R. (2016) 'The rise of uncertainty', *Critical Horizons*, 17(2), pp. 160–167.

Castells, M. (1996) *The Rise of the Network Society*. Oxford: Blackwell.

Castells, M. (2012) *Networks of Outrage and Hope: Social Movements in the Internet Age*. Cambridge: Polity Press.

Cingolani, P. (2014) *Révolutions précaires: Essais sur l'avenir de l'émancipation*. Paris: La Découverte.

Colombo, E. and Rebughini, P. (eds) (2019) *Youth and the Politics of the Present: Coping with Complexity and Ambivalence*. London: Routledge.

Crouch, C. (2000) *Post-Democracy*. Cambridge: Polity.

Diamanti, I. and Lazar, M. (2018) *Popolocrazia*. Roma: Laterza.

Farro, A.L. (2000) *Les movements sociaux: Diversité, action collective et globalisation*. Montréal: Les Presses de l'Université de Montréal.

Farro, A.L. (ed.) (2006) *Italia Alterglobal*. Milano: FrancoAngeli.

Farro, A.L. and Rebughini, P. (eds) (2008) *Europa Alterglobal*. Milano: FrancoAngeli.

Farro, A.L. and Lustiger-Thaler, H. (eds) (2014) *Reimagining Social Movements: From Collectives to Individuals*. Farnham: Ashgate.

Findlay, R. and O'Rourke, K. (2007) *Power and Plenty: Trade, War, and the World Economy in the Second Millennium*. Princeton, NJ: Princeton University Press.

Fraser, N. (2017) 'The end of progressive neoliberalism', *Dissent Magazine*, 2(1), pp. 1–8.

Galbraith, J.K. (1967) *The New Industrial State*. Princeton, NJ: Princeton University Press.

Geiselberger, H. (ed.) (2017) *The Great Regression*. Cambridge: Polity Press.

Lyotard, J.F. (1979) *La condition postmoderne:.rapport sur le savoir*. Paris: Les Éditions de Minuit.

Martuccelli, D. (2017) *La condition sociale modern: L'avenir d'une inquietude*. Gallimard: Paris.

Mazzoleni, G. and Bracciale, R. (2018) 'Socially mediated populism: the communicative strategies of political leaders on Facebook', *Palgrave Communication*, 4 (50), pp. 1–10.

Melucci, A. (1996) *Challenging Codes: Collective Action in the Information Age*. Cambridge: Cambridge University Press.

Mouffe, C. (2018) *Pour un populisme de gauche*. Paris: Albin Michel.

Mudde, C. and Rovira-Kaltwasser, C. (2017) *Populism: A Very Short introduction*. Oxford: Oxford University Press.

Rebughini, P. and Sassatelli, R. (2008) *Le nuove frontiere dei consumi*. Verona: OmbreCorte.

Rebughini, P. (2010) 'Critique and social movements: looking beyond contingency and normativity', *European Journal of Social Theory*, 13(4), pp. 459–479.

Rorty, R. (1989) *Contingency, Irony and Solidarity*. Cambridge: Cambridge University Press.

Rosa, H. and Scheurman, E. (eds) (2009) *High Speed Society: Social Acceleration, Power, and Modernity*. Philadelphia, PA: The Pennsylvania University Press.

Rosanvallon, P. (2006) *Counter-Democracy: Politics in an Age of Distrust*. Cambridge: Cambridge University Press.

Rosenau, P.M. (1992) *Post-Modernism and the Social Sciences: Insights, Inroads, and Intrusions*. Princeton NJ: Princeton University Press.

Urbinati, N. (2019) 'Political theory of populism', *Annual Review of Political Science*, 2, pp. 111–127.

Schmitt, C. (2005 [1922]) *Political Theology: Four Chapters on the Concept of Sovereignty*. Chicago: University of Chicago Press.

Touraine, A. (2007) *A New Paradigm: For Understanding Today's World*. Cambridge: Polity Press.

10 Intercultural critical reflections on postcolonialist-decolonialist and populist theories from Latin America and Ecuador

Luis Herrera-Montero and Lucía Herrera-Montero

Populism is a topic of relevance in Latin American social analysis, and, during the 1970s, it was a useful theoretical tool in the understanding of a variety of contexts and political processes that were taking place in the region in the twentieth century. This essay focuses on contemporary Ecuadorian reality. The topic is approached from the standpoint of different cultural manifestations which demonstrate the presence and effectiveness of colonial hegemonies, expressed in clientelisms and idealisations of certain political leaders which, in turn, reveal both cultural and historical complexities that, although a means of subordinating and normalising the people, also expose resistances that enable them to directly negotiate and obtain benefits in everyday life, an issue that, Ernesto Laclau (2005) addressed in his text *On Populist Reason*. In response to such a concern, we suggest the rescuing of postcolonial theoretical standpoints. However, the basic argument of this essay arises from the hypothesis that it is essential to conceive intercultural projects, which entail not only decolonised consciences, but also the begetting of new civilizational pacts. In such a context, political culture is supported by ethical concerns and parameters, as well as by the belief that it is necessary to share social and political power. From this perspective, interculturality goes beyond postcolonial theories, socialist approaches and, even, its own indigenist stances.

To start with, this essay reflects on populism as a colonial manifestation of political endeavour. Subsequently, it explores populist behaviours from the standpoint of postcolonial analysis. Finally, it proposes interculturality as a means of critique of postcolonial deficiencies and inadequacies, as well as of the indigenist shortages of certain theoretical trends in contemporary Ecuador.

Introduction

The Ecuadorian left wing longs for social change, but does not have enough awareness of its own vulnerabilities when facing the pervasive hegemony of capitalism. As a result, revolutionary movements have not had clear strategies to develop sustainability and skills of reproduction with incidence levels equivalent to those of the hegemonic and dynamic practices of capitalism. Nonetheless, our intention here is not to contradict institutional processes of the left-wing linked regimes in the country and in the region. Neither do we pretend to evade or invisibilise processes of resistance to the classist, oligarchic and colonialist management of social power. On the contrary, our work entails analysing studies on populism and postcoloniality-decoloniality, as trends of political interpretation which have had a significant resonance in the last three decades in the region, as they have characterised the epistemic foundations of progressive governments and of social movements, respectively. However, we also intend to point out their insufficiency for concerted political action in our country and, perhaps, in Latin America as a whole. The effort that we share here does not involve an attack upon, or a defence of, this or that specific postulate, but an exploration of its scope and limitations when dealing with the hegemonic conditions and regarding the execution of processes towards social change. In this endeavour, we consider it imperative to carry out an exercise of dialogue that, grounded on the interculturality perspective – which features epistemic-epistemological and political contributions and which constitutes a hallmark of Latin American thought – will enable the analysis and criticism both of postcoloniality-decoloniality and of populism.

Our text begins by problematising our political crises, as they may constitute the prelude to the attainment of any civilisational change. However, given the crisis situation in our region and after the establishment of new regimes, the problem unvaryingly persists due to an inadequate understanding of the hegemonic practices of capitalism: the system responds dynamically in order to ensure its own and continuous reproduction by readapting any process within its hegemonic structure. In such a context, postcolonial-decolonial and populist proposals show an analytical shortage that prevents them from gaining an accurate understanding of these schemes of assimilation and recomposition within capitalist hegemony.

The second part of the text exposes succinctly our work methodology. The section is mainly a review of authors' contributions on the ideas of rhizome and hermeneutics. It is not a state-of-the art analysis of postcoloniality-decoloniality, populism or interculturality, as we have chosen authors who have a greater analytical positioning in their respective areas of study and who have explained approaches to hegemony and resistance practices.

In the third part of this text, we present a critical account of the populist and postcolonial-decolonial approaches from the perspective and postulates of Latin American interculturality. Concerning this last theoretical axis, our interest is to develop the topic from the standpoint of proposals that promote dialogic exchanges between notions of identity, difference and recognition of epistemic plurality; that emphasise the need to build critical reflections, and that prioritise theoretical paths rooted in transformative praxis.

Problematisation of the current political crisis, as a prelude to understanding the feasibility of postcolonial-decolonial and populist proposals

The social and economic deterioration of living conditions in Latin America, due to neoliberalism relapse, have produced a generalised state of crisis, from which Ecuador did not escape. The situation that our country experienced at the turn of the century is a good example of what was happening in the region: in 1999, the Ecuadorian government implemented several policies in order to rescue a number of important banks from bankruptcy: a bank holiday that froze deposits of the population in the financial system was put into effect, whilst our former currency (sucre) experienced a particularly strong devaluation. As a consequence of the financial and subsequent social crisis, massive migratory processes[1] occurred. In response to critical situations such as ours, proposals for social transformation emerged worldwide; the World Social Forum was established as a confluency scenario for parties, social movements, non-governmental organisations and even states, through the involvement of several progressive Latin American governments. Various nations coupled in their own political projects several postulates from the World Social Forum, as evidenced by the contents of new constitutions and action plans of both national and local governments. The Forum's main motto was to position the political struggle into the idea that "Another world is possible".

Despite the broad irradiation of such ideas, their sustainability was insufficient given the upholding of a display of privileges that were easily assimilated by the global hegemony of capitalism. As a result, local governments also ended in attrition and were severely questioned. In Ecuador, two clear examples were Cotacachi and Guamote, both local governments which, for approximately fifteen years, had been considered as an emblematic result of successful processes of social organisation. These sorts of problems were replicated at a national level by the progressive and neo-populist governments which had emerged in Latin America as political alternatives in the course of the neoliberal crisis of the late twentieth century and early twenty-first. Regarding Ecuador, such a process began in 2007, with the electoral victory of Rafael Correa, and ended up several years afterward provoking serious criticism and disappointment as it became obvious that, besides there being little or no rotation of political leadership, there also existed a substantial inherence of the state in social organisation roles, excessive levels of authoritarianism and recurrent corruption accusations.[2]

When faced with these numerous questionings, processes of self-criticism did not prevail in Rafael Correa's administration, and capitalism was accused as the only and great culprit of the deterioration and detainment of the revolutionary processes in the region. However, even if this aspect is also true and obviously undeniable, the capitalist onslaught cannot be considered an excuse to carry out evaluations out of rigor regarding mistakes and deviations of the political project of civilising change, established in the Constitution of

2008. Given this situation, the transformation processes were finally neutralised. At the moment, there are, on the one hand, social movements focused on strictly local experiences, with obvious weaknesses in national scenarios; and, on the other hand, progressive governments in an acute state of weakness before a global economic trans-nationalisation of political action.

Crises can produce alternative processes of change, but can also provoke disappointment and reactivate practices that contradict consciences and emotivities aimed for the common good. These reactive practices generally entail tendencies toward disrespect, violence and anger, and are prone to reproduce authoritarianisms and totalitarianisms. Examples are visible in the rise of neofascist tendencies in Colombia and Brazil, as the crises of the capitalist system impose reactive manifestations of politics that disrupt the possibilities of collective-societal transformations and degenerate survival through highly destructive structures. Within the actual state of things, the capitalist hegemony has been able to regenerate and expand itself globally, regionally, nationally and locally: thereupon the manifest rise of rulers with neoliberal affiliations in the region, and the formation of PROSUR – the bloc to which Ecuador entered with our current president, Lenin Moreno – as an international alliance opposed to UNASUR. At the same time, the interconnected responses of the anti-hegemonic proposals in the region appear extremely weak and highly vulnerable to the dynamics of the system's hegemony.

Given this context, we consider it imperative to analyse whether the theoretical and epistemic referents, which accompanied these struggles for social change, are also in crisis: what limitations are there in the analysis of the hegemonic conformation of capitalism? Of course, we do not intend to reactivate socialist, communist or anarchist proposals of former centuries; but we do wish to consider the analytical effectiveness of two current theoretical trends, postcoloniality-decoloniality and populism, from the viewpoint of an eminently Latin American epistemic reference: interculturality, which can address gaps and establishes dialogues with undeniably valid aspects of both theories.

Rhizome and hermeneutics in the interpretation and understanding of texts

The validity of dialectical materialism could be sustained on its methodological contribution to analytical perspectives and theoretical constructions that contrast with philosophical idealism and scientific positivism.[3] Correspondingly, Marx (1974) highlights the dialectical relationship between subject and object in the epistemological and socio-political analysis of reality. He questions the objective negation of the real, as well as the materialistic mechanism of annulling processes of rationality and protagonism of the subject, both in the epistemological and socio-political aspects.

Nevertheless, based on Deleuze's critique of Hegel (1983), we disagree with the dialectics. Deleuze aligns himself with Nietzsche as he considers that the becoming does not proceed from the idea but is the product of the will. Along these lines, we find a correlation with the Marxist concept of praxis, provided that the subject is not only a conscious actor capable of transforming reality, as the action is predominantly expressed within desire, instinct and the unconscious. In this sense, the transformative becoming is not an act of practical rationality, but primarily an act of will-vitality. A second discrepancy with the dialectics lies in the meaning of denial. From the Hegelian perspective, the difference appears always as antithesis or denial, when it must be conceived as an affirmation (Hardt, 2004). In Hegel's dialectics, difference emerges subordinated to identity, and, therefore, susceptible to reactive actions that prevent its unfolding in the becoming (Deleuze, 2002). Finally, the third point of disagreement is in the concept of unity and synthesis. Hegel, as well as Marx, favours the synthesis of multiple determinations and the unity of the diverse (Marx, 1982). On the contrary, the main characteristic of becoming is that it never abandons its condition of multiplicity-plurality. For Deleuze, and for Nietzsche, reality acquires a sense of difference as soon as it is woven and flights in multiplicities.

> Long live the multiple! although it is now very difficult to launch that cry. No typographical, lexical, or even syntactic ability will be enough to make it heard. The multiple must be done, but not constantly adding a higher dimension, but, on the contrary, in the simplest way, by force of sobriety, at the level of the available dimensions.
>
> (Deleuze and Guattari, 2005, p. 12)

For its better understanding, the rhizome is explained by means of the following principles: (1) and (2) connection and heterogeneity, understood as links – whether they be political, linguistic, economic, or semiotic – that account for singularities; (3) multiplicity: the rhizome opposes the one, and has no subject or object, but dimensions, relationships, determinations; (4) a-signifying rupture: a rhizome breaks anywhere, but reshapes itself into another; it has lines of segmentarity and lines of flight; processes of deterritorialisation and reterritorialisation; (5) cartography, which is based on maps and not on decals, and differs from deep and tree structures, as well as from dual conceptions – understood mainly from moralities marked by borders between the good and the bad. With the rhizome, maps are drawn in a variety of instances, be it walls, works of art, and books, among others. Whereas decals, photographs or radiographies, are characterised by the arboreal; by structuring the rhizome, blocking the exits, neutralising multiplicities and reproducing themselves. In the context of our analysis, a decal characterises the capitalist society: it limits the maps, prevents lines of flight, submits the value of the difference, functionalises everything within the capital-work structure, which has the ability to change, reproducing itself constantly, preventing the free becoming.

A rhizome as an underground stem differs radically from roots and smaller roots. The bulbs, the tubers, are rhizomes. But there are plants with roots or smaller roots that from other points of view can also be considered rhizomorphs. One might ask, then, whether botany, in its specificity, is not entirely rhizomorph. Even animals are when they go in a pack, rats are rhizomes. The burrows are so in all their functions of habitat, provision, displacement, of den and of rupture.

(Deleuze and Guattari, 2005, p. 12)

The rhizome, as Deleuze and Guattari define it, can be evidenced in the book, within which articulation lines, segmentarity, strata and territorialities are noted, as well as lines of flight, deterritorialisation and unstratification movements: "there is a parallel evolution of the world and the book, the book ensures the deterritorialization of the world, but the world makes a reterritorialization of the book, which in turn deterritorializes in itself in the world" (Deleuze and Guattari, 2005, p. 16). On the basis of associating the rhizome with the book, we carry out our understanding of theoretical arguments. Our analysis is not based on studying factual processes of the real, but on addressing theories, through which Latin America's reality has been scrutinised.

Along with the concept of the rhizome, we address hermeneutics as a methodology of textual interpretation. We base ourselves on the contributions of Gadamer (1977), for whom all language and all knowledge are first and foremost interpretation. The act of theorising grants value to the textual interpretation; as Bauman (1978) asserts it, truth is constructed from interpretations and understandings in constant transformation. Scientific truth, therefore, has an unavoidable connection with communicational consensus, and is not an uncontested, absolute and universal condition. Bauman (1978) retakes contributions from Heidegger regarding the assertion that explanations about existence and the world, depend on how human beings interpret their existence from being and being in the world.

In an effort to integrate our two main methodological perspectives, we suggest a procedure based on what could be called a hermeneutical rhizome. With that in mind, we organise our critique in two correlated units: (1) the review of theoretical contributions on postcoloniality-decoloniality and populism in Latin America; and (2) the exposition of a critical approach that, based on intercultural epistemic guidelines, fosters simultaneous dialogues among identities and differences around theorical and political topics. On the topic of postcoloniality and decoloniality, our review deals with the academic perspectives of authors like Edward Said (1978) and Ranajit Guha (2002), as precursors of the postcolonial debate on South Asia, and, in the Latin America academic context, with authors such as Walter Mignolo (2007), Enrique Dussel (2009), and Aníbal Quijano (2007). As for the studies on populism, we tackle the works of Torcuato Di Tella (1973), Francisco Weffort (1998), Octavio Ianni (1973) and, most significantly, Ernesto Laclau (2005). On interculturality, we discuss Catherine Walsh (2007), who has understood interculturality under the paradigm of the decolonial turn, and Alba and Ruth Moya (2004), Fidel Tubino (2005), and Raúl Fornet-Betancourt (2007), whose approaches share positions that diverge from the decolonial aim of exclusivity regarding sociocultural problematics.

Criticism of postcoloniality-decoloniality and populism from the standpoint of an intercultural episteme

A proper understanding of postcolonial and populist proposals must start, in our opinion, with a clear conceptualisation of the hegemonic practices, with the aim of elucidating the complex mixtures and leaks between ideological domination and emancipation. Hegemony, as defined by Gramsci (1981), consists in the cultural control that dominant classes have over the general consciousness; that is, the ideological supremacy or the ability to determine the political sense of society. However, also within hegemony, Gramsci conceived new insights and understandings, namely, a new ideology which aims to replace the old political system. He understood the core contents of the hegemonic construction as a complexity of processes and contexts, both continuous and discontinuous.

> The necessary work is complex and must be articulated and graduated: there must be combined deduction and induction, identification and distinction, positive demonstration and destruction of the old. Not in the abstract, but in the concrete: based on the real [...]: changes in the ways of thinking, in beliefs, in opinions, do not happen due to rapid and generalized "explosions", they commonly occur by "successive combinations" according to extremely varied "formulas".
>
> (Gramsci, 1981, p. 100)

Gramsci's arguments are crucial when pondering that social change involves disputes and sustained processes in the real. He is utterly aware that transformation implies a diversity of contexts, actors and practices. Nonetheless, the problem in his perspective is his attempt to encompass the multiple within a synthetic pathway; that is, according to his Marxist training, within a new consciousness or legitimate exercise of hegemony. Our explanation of the phenomena does not deny Marxist-Gramscians' stances, as in their work it is already possible to discern theoretical contents that understand the real in terms of its complexities and diversities. From a comparable standpoint, we assume difference as the construction, deconstruction and reconstruction from the expansive and rhizomatic multiplicity.

Along this line of reflection, we attempt the explanation of postcolonialism and populism mainly as a theoretical exercise linked to processes of becoming and perishing, as acts of continuing and discontinuing. As we have already stated, our theoretical premises differ from the dialectical understanding which, as Engels (1974) points out in his critique on Feuerbach, comprises a progressive-evolutionary perspective, according to which whatever is superior becomes whilst that which is inferior perishes. From the standpoint of Deleuze's rhizomatic understanding (Deleuze and Guattari, 2005), there is "a truly crazy-becoming", where the younger becomes older than the old, as the older becomes younger than the younger; a becoming that articulates past and future.

We formulate the concretions of hegemony as definitely multiple, since they refer to complex and diverse domination practices. The problem is that such domination is not only verified, but also normalised in a variety of contexts. Thereby, a plurality of peoples and populations tend to conceive themselves as inferior, normalising themselves as subalterns. In Gramsci's words, this responds to the fact that hegemonising acts must be understood as weaves whose threads are heterogeneous and become concrete in multifaceted ways; thus, one can be hegemonic as a bourgeois, man, father, bureaucrat, priest, and so on; but also, as a woman, a worker, a child and a homosexual. Once in reality and in people's everyday life, regarding the radical heterogeneity of the hegemonic weaving, domination and subalternity can be experienced simultaneously: even being a subaltern one can exert hegemony over others who, in turn, see themselves as inferior. Indeed, colonialism has been put in place on the basis of this complexity of hegemony: it merges the alleged superiority of class, ethnicity, gender, age, among others, in the form of multiple interdependencies and scales of socio-cultural hierarchy, where, nonetheless, what appears to be new can be extremely old or, in fact, actually new. In the colonial hegemonic processes, there is no real fracture, but a discontinuous continuity and a continuous discontinuity: "truly crazy becoming", as Deleuze declares.

The colonial complexity acknowledges the dynamism of domination and hegemony practices. It is certainly a dynamic process which, across more than 500 years, has maintained systems and regimes based on a purported racial, epistemic and linguistic superiority of Western society, and it has so done according to the evolution of the capitalist system in its mercantilist, industrial and scientific-technical versions. Hence, considering the decolonial-postcolonial ideas as trends that have lost relevance is a crass error seized by certain theoretical currents that have only modernised the colonial normalisation.

Postcoloniality and decoloniality are theoretical streams that emerged from different contexts; however, it is not misbegotten to assume that decoloniality emerged from originally postcolonial bearings. Postcolonial studies arose from the intellectual work of various disciplines which focused on the analysis of Western domination and colonisation around the end of the 1970s.[4] Edward Said (1978), in *Orientalism*, questions the deceptiveness of supposedly oriental stances that, in fact, are Western interpretations of an unreal East. Ranajit Guha (2002) also denounces the unreliable colonial approaches to India's history that, responding in this case to nationalist and Marxist outlooks, did not foster the generation of a historiography particular to the South Asian reality itself. Walter Mignolo (1995), a Latin American intellectual, makes epistemic and hermeneutical precisions on the topic of postcolonialism. He establishes a clear difference between the postcolonial situation and the postcolonial theorisation. The coexistence between becoming and perishing that characterises all theoretical-practical processes is being highlighted: while postcoloniality involves the overcoming of colonial conditions as well as the subsistence and continuance of colonial elements, the decolonial turn highlights the becoming, as rupture and flight, matching in this way the notion of the rhizome. On the basis of this overall explanation, it becomes comprehensible that postcoloniality involves theoretical postulates which show clear influences from the poststructural thinkers – i.e. Derrida, Foucault and Lacan – in the aforementioned postcolonial intellectuals.

While focusing on Latin America, let us dwell on Mignolo's (1995) arguments in order to elucidate the unveiling accounts of Western geopolitical and geohistorical impositions, which the author defines as colonialism. Mignolo stresses that, in historical terms, postcoloniality can be associated with independence processes of former colonies, emphasising as well that elements of coloniality remain in any postcolonial situation. In this regard, there is a difference between Latin American and Caribbean countries' postcoloniality – especially Jamaica – and the United States'. In this order of ideas, the author distinguishes two different ways of understanding postcoloniality: as a characteristic of a specific historical situation and as a product of theoretical approaches. That is to say, to affirm the presence of postcolonial situations in the independence processes does not carry the same meaning of the 'post' in the theoretical production: this second notion involves geocultural contents that could serve as guidelines for theories that advocate on legal, political, historical and literary emancipation.

Decoloniality, meanwhile, is strictly identified as a Latin American stance, which, from the year 2000 onwards, can be found in the theoretical production of academics like, again, Walter Mignolo (2007), Enrique Dussel (2007) and Aníbal Quijano (2007), who aim to overcome the long history of colonialism that prevails in Latin America. Decolonial epistemic guidelines focus on questioning the deceitful supremacy of the Western world over other peoples, an aspect that has undeniable contemporaneity in discourses and stratifications. In epistemic terms, all decolonial authors emphasise the hegemonic role of the West, evidenced in the dominant perspective of history, rationality and philosophical-scientific productions, among the most important.

Returning once again to Mignolo (2007), decolonial stances underline that modernity is eminently colonialist, given that it involves pervasive domination exercises as result of a fictitious social superiority of the West over the conquered peoples in America, Asia and even within Europe. Decolonial thought arises, then, as an opposing counterpart to civilising modernisation. Among these analytical lines, the decolonial turn differs from postcolonial studies insofar as it does not emerge from a poststructural genealogy. Accordingly, Enrique Dussel (2007) considers imperative the pursuit of a different thought, based on new grounds of reflection, that will not reduce the epistemic contributions of colonised peoples and societies to mere practices of imitation or to marginal comments in European or North American philosophical treaties. That is why Dussel highlights the diverse conformation of peoples and ethnic groups in the context of the postcolonial State and the legitimacy of positions like those of the Zapatista movement, as an example of "indigenous politics that is not assimilationist" (Dussel, 2007, p. 556). A significant contribution of the proposed decolonial turn focuses on recognising Western coloniality within the different stages of the unfolding of capitalism. Seen from this perspective, coloniality is clearly linked to changes among mercantilist, industrial and global domination; contemporary coloniality is indisputably global.

From Quijano's perspective, modernity imposed its civilising processes not only in geopolitical terms, but also in biological, epistemological and cultural terms. Given such an account, modernity is incomprehensible without its colonialist condition. He points out Western hegemonic practices which coexist with modern sociocultural positioning and result in constituting a Eurocentric rationality as the main axis of domination. From this constitutive basis, Western civilisation imposed itself as the mandatory future for any social process, through unidimensional perspectives of evolutionary-civilising development. A Western model of making politics and producing knowledge was instituted as a determining and universal condition both of democracy and epistemology.

In this order of ideas, Europe and the Europeans were the most advanced moment and level in the linear, unidirectional and continuous path of the species. Thus, along with this idea, another of the main nuclei of Eurocentric modernity/coloniality was consolidated: a conception of humanity, according to which there is a differentiation among the population of the world: between inferior and superior, irrational and rational, primitive and civilized, traditional and modern.

(Quijano, 2007, p. 95)

Quijano's criticism of the supremacy of modernity as hegemony, which subordinates other cultures and epistemes, encompasses Marxist thought. Within this particular questioning is the rebuke of the exclusivity of scientific rationality as a reference point for understanding and explaining social reality. Marxist thought devalued, as part of what was considered ideology, any other form of interpretation of the world, production of knowledge, and building of social and intersubjective relationships.

In the context of Ecuadorian academic production, Catherine Walsh (2007) has undertaken the decolonial discussion intertwined with interculturality paradigms, distinctive of Latin America.

The concept of interculturality has a meaning in Latin America, and particularly in Ecuador, linked to place and space geopolitics, from the historical and current resistance of indigenous people and blacks, to their constructions of a

social, cultural, political, ethical and epistemic project aimed towards decolo-
nization and transformation.

(p. 43)

Still, it should be made clear that interculturality is not a derivation of the
decolonial turn. It is crucial to elucidate this epistemic distinction since the
sources of the Ecuadorian indigenous movement predate the theoretical
trends of decoloniality. Already in 1945 and long before the decolonial focus,
the indigenous movement in Ecuador started its struggle, specifically in the
sphere of education. Furthermore, interculturality includes and articulates
within its paradigm epistemic and political perspectives beyond decoloniality.
Therefore, it is nonetheless valid to understand, as Walsh points out, interculturality as a framework that fosters epistemic dialogs between peoples
from indigenous, Western or other latitudes descent.

In Ecuador, contributions such as those of the sisters Alba and Ruth Moya
(2004) expose the undeniable influences of Marxist paradigms on the intercultural paradigms of indigenous movements. Along the same line, Fidel
Tubino (2005) states that indigenous peoples are not against modernity and
democratic perspectives, though certainly in defiance of an exclusionary
modernity. Tubino's account does not display the anti-modernisation arguments we acknowledged as important traits of the decolonial turn. Though
the need for decolonisation is obviously not dismissed, the intercultural proposal is fundamentally dialogical and inclusive, which means that cultures of
varied origin, including, obviously, the Western one, are not underestimated
and even less despised. For Fornet-Betancourt (2007), interculturality is a
philosophical tradition that learned to recognise the plurality of cultures that
live in America. He points out that interculturality emerged from indigenous
peoples, but it is not limited to its epistemic tradition, and therefore it is
willing to embrace other contents and perspectives. Along these lines, Latin
American philosophy, for instance, seizes the insights to recognise and nurture
itself from the epistemic plurality of interculturality, and the power-sharing
community reciprocity relationships. According to interculturality criteria,
one must even unlearn the decolonial subalternity, which entails the existence
of a dominating counterpart.

Nevertheless, ignoring the contributions of the decolonial turn is not the
aim of this work. Denying the existence of colonisation and of the alleged
supremacy of the West, whose impacts were extremely denigrating and pro-
moters of severe processes of cultural, linguistic and epistemic extinction,
would be simply absurd. A biased estrangement will only reproduce the prejudices of Western universality, which must be rigorously questioned for
imposing philosophical and scientific rationality as superior to any other
episteme. It is imperative, without a doubt, to decolonise. However, the aim
cannot be to homogenise the West, modernity, Marxism and post-
structuralism; by doing so, we will be only reproducing the practices of capitalist hegemony that we have strongly criticised. At this point, Spivak's (1999)

argument, which sustains, as an epistemological right, that need to assume ourselves to be Westerners as well, turns out to be useful. The decolonial debate cannot simply disregard Western epistemology. Perspectives such as those of Mignolo, Dussel, Quijano and Walsh should recognise that Western thinking is already present in their own decolonial paradigm; without concepts such as Gramscian hegemony or Foucauldian episteme, for example, decolonial theories would simply be ignoring an important part of their own epistemological trajectory.

Our analysis on coloniality, as well as the one we undertake on populism, does not dismiss instances where hegemony also implies resistances to systems of capitalist domination. Accordingly, populism also entails processes of continuity and simultaneous discontinuity of domination and emancipation realities. We found it necessary to briefly address trends that reduce the concept of populism to some sort of political degradation, where the leading populist actors are foremost players without clear projects in the complex development of the correlation of national and transnational forces, as is the case of several Ecuadorian studies on the field. However, we estrange ourselves from such perspectives and take, as a basis for our study, the arguments of Laclau (2005).

Torcuato di Tela (1973) points out that populism cannot be reduced to underdevelopment contexts and to masses with a low level of education. It should be understood as atypical opposition to the status quo, of people not linked to the working class and its experience of social struggle, nor to middle-class sectors which are identified with technical languages outside popular sectors. Populism arises from crises of democratic and authoritarian systems, which characterised different governments in Latin America at the end of the nineteenth century and the first decades of the twentieth century. Though it emerges in socioeconomic conditions of underdevelopment contexts, populism responds to a synergy between political leaders and popular sectors in Latin America. As a consequence, for Di Tela populism is eminently plural: ergo, populisms can align themselves and be defined as conservative, moderate and radical, depending on the sector and context in which they are constituted. Francisco Weffort (1998) agrees that populism is a highly heterogeneous subject, a feature that makes it difficult to provide a rigorous and clear-cut conceptualisation. In an effort to construct a succinct overview of the populist phenomenon, Octavio Ianni defines it as "a mass movement that appears in the center of the structural ruptures that accompany the crises of the world's capitalist system and the corresponding crises of the Latin American oligarchies" (Ianni, 1973, p. 85).

From a different perspective, Enrique Dussel (2007) understands populism as a process of domination or hegemonic exercise carried out by the industrial bourgeoisie – not necessarily national – in conflict with the hegemony either of the agrarian landowning oligarchy or of a dependent mercantilist segment; which, in turn, correspond to two different stages of development of the capitalist mode of production. In this context, according to Dussel, populism thrusts the peasantry to oppose the landowning oligarchy and give way to the industrial mode of production. Another characteristic that Dussel highlights of populism is that it frames an alliance between marginal sectors and political leaders, given the demographic growth due to emigration from the countryside to the city and the precarious conditions that such displacements bring about. The cause for such a situation lies in the incipient development of the emerging industrial capitalism which cannot absorb the magnitude of social demand for employment. Finally, Dussel classifies the populist state as peripheral capitalist, which distinguishes it from the popular socialist proposal of Cuba and the fascism of Hitler and Mussolini. In short, Dussel does not find in populism any token of processes of resistance.

In Ecuador, among the most relevant inquiries on populism, are those of Agustín Cueva (1988), Rafael Quintero (1980), Pablo Cuvi (1977), and Amparo Menendez (1986). The references we make here to such studies correspond to a compilation of texts presented by Felipe Burbano de Lara and Carlos de la Torre (1989), which include many of the aforementioned authors, among others. The analysis set forth in its introduction starts by agreeing that there is no consensus on the interpretation of the concept. As in analyses made on Latin American populism, they share the opinion that the emergence of popular sectors in politics results from the crisis of the oligarchic liberal system. They highlight leaderships that have been able to draw and represent popular senses that burst onto the political scene in an atypical way, radically differing from representativeness and political action under modernity's parameters, but vulnerable to manipulation and transnational capitalist hegemony.

So far, we have corroborated that populism is not alien to coloniality. On the contrary, the fractures in the oligarchic regime and the emergence of popular sectors in the political scenario, do not imply the conformation of political structures opposed to historical processes of domination and hegemony. However, we should be careful not to downplay, as some of these authors actually do, socio-popular forms and expressions that constantly demarcate themselves from the system's normative order and, therefore, also resist coloniality. This poses the challenge of conceiving a different hermeneutic, one that enables a new understanding of populism, closer to the concept of rhizome and where it would be possible to discern maps and lines of flight.

Ernesto Laclau (2005), whose work is central in the characterisation of contemporary populism, outlines an interesting analytical procedure which highlights the need to agglutinate diverse social resistances that confront dominant hegemony and power structures: "the demand requires some kind of totalization if it is going to crystallize in something that can be inscribed as a claim within the 'system': all these contradictory and ambiguous movements imply the various forms of articulation between logic of difference and logic of equivalence" (p. 9). Laclau's thesis concurs with the intercultural perspective. By elucidating the relationship between identity and difference, intercultural thinking implies a dialog between the I and the other. This is how Fornet-Betancourt (2009) puts it: "The subject, especially understood in the terms of 'I', can be read as a construction of the human conscience whose purpose is no other than to control what happens in the fields of the aware-ness process. This means that conscience cannot be reduced to the scope or dimension of what we call 'I' or 'ego'" (p. 43).

Going back to Laclau, populism is a phenomenon that aligns itself to the whole scope of politics, whereas politics is crossed by populist dynamics. In the words of Retamozo (2017), the concept establishes a dichotomy between us–people and the others–power. Within this logic, Laclau formulates a defi-nition of populism that clearly contradicts former ones. In a certain way, it is similar to the one Di Tella's proposes; that is in the sense of populism being opposed to the status quo.

> There exists in every society a reservoir of pure *anti-status quo* feelings that crystallize in some symbols in a relatively independent way from the forms of their political articulation, and it is their presence that we per-ceive intuitively when we call "populist" a speech or a mobilization. Cli-entelism – to return to the example – is not necessarily populist [...]
>
> (Laclau, 2005, pp. 156–157)

At first glance, we could have the impression that the category of people arises in Laclau's argument as a homogeneous entity, when in fact it is extremely heterogeneous, as the rhizomatic multiplicity. The ambiguity dis-appears as he introduces and explains the concept of the empty signifier:

> The empty character of the signifiers that give unity or coherence to the popular field is not the result of any ideological or political under-development; it simply expresses the fact that all populist unification takes place in a radically heterogeneous social terrain.
>
> (Laclau, 2005, pp. 127–128)

The author makes it clear that the empty signifiers represent the symbolic need for attention to a multiplicity of unsatisfied demands that the people pose: a clear example of a-signifying rupture and lines of flight. The borders that separate the power from the people are highlighted at the moment this latter gathers together around that symbolism in the aim of a radical totalising democracy.

If we compare these postulates with the multiplicity proposed as a rhizome, we find out that, along similar lines of reflection, both the rhizome and populism privilege the affirmative value of difference, that is, not in terms of a dialectical relationship, but as an utterly plural one:

> if – given the radical heterogeneity of the links that intervene in the equivalence chain – the only source of coherent articulation is the chain as such, and if the chain only exists whilst one of its links plays a role of condensation for all others, in that case the unity of the discursive

formation is transferred from the conceptual order (logical of the difference) towards the nominal order.

(Ibid., pp. 129–130)

This quote reveals a clear proximity between Laclau's idea of populism and the paradigm of interculturality. Nonetheless, regarding interculturality, the category of the people within the perspective of indigenous cultures, just like the concept of community, always has a pluralising quality: in Latin America, it can never be assumed that there is one single and comprehensive community or a single and comprehensive indigenous people (*un pueblo originario*). Thus, interculturality invariably implies the coexistence of communities, peoples and nationalities. It is necessary, then, to pluralise the concept to the peoples (*los pueblos*), though the issue of universality is indeed a priority for the indigenous peoples, as it is for Laclau, who states that the broad presence of particularities complicates the concretion of the equivalence chain. In this sense, it is important to point out that the idea behind the "community of indigenous peoples" is of reciprocity between identities and differences. To be part of a community (*ser comunitario*) supposes a long trajectory of public service, in the line of the *communitas* proposal by Roberto Esposito (2012) and the systematisation of the political community, as elaborated by Chantal Mouffe (1999).

A difference that should be highlighted between the perspectives of interculturality and that of Laclau on populism has to do with the meaning of power, which is not necessarily the hegemonic domain of a system. In the indigenous worlds, supported by communities, power is shared. Therefore, within intercultural relations, power is learning and unlearning sustained also under reciprocal relations (Herrera, 2015). Consequently, power can be associated with structured hegemony exercises, even from empty signifiers, within the people's livelihood; that is to say, the people can resist even when reproducing symbolisms which have been structured through long processes of colonisation. There is no such obvious limit between power that dominates and power that emancipates, they are woven as the calc and the map in the rhizome.

Conclusions

Political crisis conditions are also explained through limitations in the theoretical perspectives on social change. Decoloniality should not exclude the modernisation process and the West, since it also carries features of resistance and counterhegemony, without which there would be no postcoloniality or decoloniality.

The proposal for an intercultural dialogue is more than a struggle and concretion of decolonisation processes which though important are not enough, since power is not shared from a subaltern self-assimilation position, given the connotation of inferiority that it carries.

Populism implies the existence of coloniality as it does not overcome class hegemonies, but it also comprises resistances in terms of empty signifiers and disputes between people and power.

Interculturality perspectives can enrich the idea of the people as a totality and, simultaneously, the affirmative presence of difference, as they recognise reciprocal relations among peoples and communities. From that standpoint, power, if shared, is not opposed to the people.

The methodological approach around postcoloniality-decoloniality, populisms and interculturality was elaborated through hermeneutics and the rhizome. The text was constituted on the basis of the affirmative and multifaceted condition of difference, present in the intercultural paradigm, as an episteme shared from Latin America to the world.

Notes

1 Data establishes that there were approximately 25,000,000 Ecuadorians who settled in the United States and Spain, mainly, and a smaller number in other countries such as Italy and Canada.
2 The most famous case is that of former president Jorge Glas, against whom a sentence of unlawful association was issued, when evidence emerged of negotiations by his uncle with the company ODEBRECH for approximately $US16,000,000.
3 Marx does not use the concept of scientific positivism, but in his rigorous critique of the mechanistic materialism of Feuerbach, he leaves an evident position of questioning to what in social sciences we understand as positivism. Obviously, this criticism was later made by Marxist theorists.
4 It is not possible to mention each of its members, which is why we make a brief reference to those who are considered priority sources; obviously, other authors have the same significant weight, but, we insist, it is not a state of art on a postcolonial proposal that defines this text.

References

Bauman, Z. (1978) *Hermeneutics and Social Science*. New York: Columbia University Press.

Burbano, F. and De la Torre, C. (1989) *El populismo en el Ecuador*. Quito: ILDIS.

Cueva, A. (1988) *El populismo como problema teórico y político en las democracias restringidas de América Latina*. Quito: Planeta.

Cuvi, P. (1977) *Velasco Ibarra: El Ultimo Caudillo de la Oligarquía*. Quito: Instituto de Investigaciones Económicas –Universidad Central del Ecuador.

Deleuze, G. (1983) *Nietzsche and Philosophy*. New York: Columbia University Press.

Deleuze, G. and Guattari, F. (2005) *Mil Mesetas: Capitalismo Esquizofrenia*. Valencia: Pretextos.

Deleuze, G. (2002) *Diferencia y Repetición*. Buenos Aires: Amorrortu Editores.

Gramsci, A. (1981) *Cuadernos de la cárcel*. Mexico City:Ediciones Era, S.A. de CV.

Di Tella, T. (1973) 'Democracia representativa y clases populares', in G. Germani, T. Di Tella and O. Ianni, *Populismo y contradicciones de clase en América Latina: Serie Popular Era*. Mexico City: Ediciones Era, pp. 38–82

Dussel, E. (2007) *Política de la liberación. Historia mundial y crítica*. Madrid: Editorial Trotta, S.A.

Esposito, R. (2012) *Communitas: Origen y destino de la comunidad*. Buenos Aires: Amorrortu Editores.

Fornet-Betancourt, R. (2007) 'La filosofía intercultural desde una perspectiva latinoamericana', *Revista Solar*, 3(3), pp. 23–40.

Fornet-Betancourt, R. (2009). *Interculturalidad en procesos de subjetivización.* Mexico City: Coordinación General de Educación Intercultural y Bilingüe.

Gadamer, H. (1977) *Philosophical hermeneutics.* Los Angeles: University of California Press.

Guha, R. (2002) *Las voces de la historia: Y otros estudios subalternos.* Barcelona: Crítica.

Herrera, L. (2015) 'El cosmopolitanismo y la interculturalidad. Un análisis desde una contribución teorico-crítica', *Revista Realis*, 5(1), pp. 203–221.

Ianni, O. (1973) 'Populismo y relaciones de clase', in G. Germani, T. Di Tella and O. Ianni (eds) *Populismo y contradicciones de clase en América Latina. Serie Popular Era.* Mexico City: Ediciones Era, pp. 83–156.

Laclau, E. (2005) *La razón populista.* Mexico City: Fondo de Cultura Económica.

Marx, K. (1974) *Las tesis sobre Feuerbach.* Moscú: Editorial Progreso.

Marx, K. (1982) *Introducción general a la crítica de la economía política.* Mexico City: Siglo XXI Editores.

Menéndez, A. (1986) *La Conquista del Voto en el Ecuador: de Velasco a Roídos.* Quito: Corporación Editora Nacional.

Mignolo, W. (1995) 'La razón postcolonial: herencias coloniales y teorías postcoloniales', *Revista CEHELERIS*, 4(5), pp. 265–290.

Mignolo, W. (2007) 'El pensamiento decolonial: desprendimiento y aperture: Un manifiesto', in S.C. Gómez and R. Grosfoguel (eds) *El giro decolonial Reflexiones para una diversidad epistémica más allá del capitalismo global.* Bogotá: Siglo del Hombre Editores, pp. 25–46.

Mouffe, C. (1999). *El retorno de lo politico: Comunidad, ciudadanía, pluralismo, democracia radical.* Barcelona: PAIDOS.

Moya, A. and Moya, R. (2004) *Derivas de la interculturalidad; procesos y desafíos en América Latina.* Quito: CAFOLIS-FUNADES.

Quijano, A. (2007) 'Colonialidad del poder y clasificación social', in S.C. Gómez and R. Grosfoguel (eds) *El giro decolonial Reflexiones para una diversidad epistémica más allá del capitalismo global.* Bogotá: Siglo del Hombre Editores, pp. 93–126.

Quintero, R. (1980) *El Mito del Populismo en el Ecuador.* Quito: FLACSO.

Retamozo, M. (2017) 'The political theory of populism: uses and debates in Latin America in post-foundational perspective', *Revista Latinoamérica*, 64, pp. 125–151

Said, E. (1978) *Orientalism.* Toronto: Random House.

Spivak, G. (1999) *A Critique of Poscolonial Reason: Toward a History of the Vanishing Present.* Boston, MA: Harvard University Press

Tubino, F. (2005) La interculturalidad crítica como proyecto ético-político. Retrieved from: https://oala.villanova.edu/congresos/educacion/lima-ponen-02.html

Walsh, C. (2007) 'Interculturalidad y colonialidad del poder: Un pensamiento y posicionamiento "otro" desde la diferencia colonial', in S.C. Gómez and R. Grosfoguel (eds) *El giro decolonial Reflexiones para una diversidad epistémica más allá del capitalismo global* (pp. 25–46). Bogotá: Siglo del Hombre Editores.

Weffort, F. (1998). 'El populismo en la política brasileña', in M. Mackinnone and M. Petrone (eds) *Populismo y neopopulismo en América Latina: El problema de la cenicienta (pp.* 135–152). Buenos Aires: Eudeba.

11 Populism

The highest stage of neoliberalism of the twenty-first century?

Adrián Scribano

Introduction

This chapter intends to answer a question that is at once both simple and complex: what does populism mean today? It seeks, on the one hand, to define a slippery concept, and to discuss/deny the existence of a populism of the left and another of the right. Populism is at the same time a political regime, a politics of sensibilities and a macroeconomic perspective for crisis management of capitalism. As set out in the Introduction to the book, the argumentative strategy of this chapter is directly connected with two tasks that are carried out allowing us to better understand what has been presented here: a conceptual approach to populism and the connection of populism with coloniality.

For all the people who have benefited from some type of populist politics, the term and experience are loaded with emotional ambivalence: the memory of the feeling of well-being and the sadness that arises from the certainty about the continuity of the causes that gave rise to the aforementioned politics.

In this chapter I try to clarify some of the reasons why an enquiry into populism as the highest stage of neoliberalism confronts us with a complex and contradictory politics of sensibilities.

In this context it is important to start with a synthetic disambiguation of the term, at least from the perspective of this text, noting that: a) populism is one of the suturing instances of capitalist crises; b) it is a politics of sensibilities oriented towards reducing social conflict; and c) it is a political regime that seeks to enthral citizenship and install sacrifice as an instance of participation.

a   Populism is a set of macroeconomic measures and social policies that enables the correction of the failures of the market and the state in its functions of reproduction of the capitalist system of social relations. In a macro-economic sense, the central focus is to increase the domestic market, favouring consumption as the engine of local production and increasing wages. From the social polices there are elaborated strategies that are articulated to the increase in consumption that the economic measures propitiate, and that serve as a guarantee of stability for the profits of some capitalist sectors.

b   Populism in its different forms of compensation (see below) is a mechanism of social pacification that contains inter-class disputes and shapes intra-class disputes through a politics of sensibilities. Populism is a practice of feeling whose aim is to give different classes a set of perceptions, sensations and emotions that explain the real. The loser is the obverse (supportive and complementary) of the winner. From its earliest forms to the present, populism is the story of the passage from well-being to well-feeling or feeling good. In this way, it emphasises and re-signifies what is experienced in the material conditions of existence, decreasing its conflictivity.

c   In the vein of b) populism consecrates sacrifice and spectacle as forms of civic participation. From the old mass management, passing through mass-media utilisation and arriving at our contemporary social networks, populism is a form of spectacle. It is a way of experiencing everyday life as a dialectic between phantoms and fantasies. The fantasy of a future world made better through the sacrifice of the present, based on the acceptance of the situation and the enjoyment of compensations through consumption. The phantom of repeating the past of privations and needs is threatened, if the fantasy of the future is not accepted. The future comes only if there is some sacrifice for the good of all, the sacredness of 'now' hides the unreality of tomorrow (Scribano 2017).

If axes a, b, and c are related, it is possible to observe how populism proposes capitalism as a cultic economy that enshrines a normalised society in the immediate enjoyment through consumption. Its political economy of morality is based on the banalisation of good, the logic of waste and the politics of perversion (Scribano, 2017, 2018).

It is in the aforementioned context that in the present chapter when I refer to populism I do not make the distinction between one of the left and one of the right, trying to argue instead that both are a dialectical overcoming of neoliberalism. In other words, it will be argued that the populism of the Global North and of the Global South are based on the same motivation: reproducing the capitalist system of accumulation and exploitation. It is within the framework of the argument that all forms of populism are colonial and therefore depredatory in an intersectional direction.

Today one of the most accepted populist characteristics has been modified: the place of patrimonial practices of power. Many authors use the patrimonial feature of rural society as the central axis to describe the "struggle" of populist movements. But in the current model of populism, the power of the owners, the regional chief and the patriarchal leaders is a key to the central power. Conversely to what happened in other stages of its constitution, populism as the highest stage of neoliberalism implies the patrimonial enrichment of the leaders, the arrival of entrepreneurs in power and the transformation into entrepreneurs of politicians. One of the most recurrent topics in the last decade around the world, both in emerging countries and those that make up the OECD, is the connection between corruption, market and institutional policy. In this context, the nation-state operates regularly on the monopoly of a populist discourse oriented to the formation of a shared identity.

Neoliberalism as populist background

Capitalist crisis and populism

Today, populism is a politics of sensibilities that continues and "improves" the results of neoliberalism. It is in this vein that populism is the highest stage of neoliberalism. When Lenin coined this metaphor to explain the relationship between capitalism and imperialism, he made a discovery: the tendency to massification and expansion of the unperceived power of capitalism shows a political and material rot of a global system of dominance. The narrative of populism maintains that its objective is to modify society, but what is pragmatically active is that behind and below the discourse of the national and popular an individualist society and self-centred consumers are built.

The situation of tension and overlap between subsidy, compensatory consumption and lumpen democracy inaugurates a stage of overcoming neoliberalism via populism. This phenomenon in the context of the emergence and expansion of populisms of the so-called left and right populisms shows that, contrary to what many have argued, populism is not the crisis, but the *result* of a crisis. Each form of populism overlaps with a type of crisis and in turn the application of measures related to neoliberalism are what cause the so-called populist forms. Obama is the solution to Bush (USA), Duhalde to De la Rua (Argentina), Evo Morales to Eduardo Rodriguez Veltzé (Bolivia), the alliance of Lega Nord and Cinque Stelle to Renzi (Italy), Chavez to the "pact of fixed point" (Venezuela), and so on.

From diverse perspectives, and with ten years elapsing between Carlos de la Torre and Jorge Savarino, they coincide in the association between crisis and populism:

> Processes of urbanization, industrialization, and a generalized crisis of paternal authority allowed populist leaders to emerge.
>
> (de la Torre, 2017, p. 196)

> The populism of the years 1930–1950 was driven by the economic crisis and the subsequent world war; today it is globalization that drives the new populist wave in Europe and Latin America.
>
> (Savarino, 2006, p. 90)

There have been particular modifications in the pattern of accumulation, models of labour management and policies of the bodies/emotions that produce a structural break and that have been the scenario for the emergence of various types of populisms.

A good example of what we intend to show is how Rosalvina Otálora Cortés (2010) has delineated the "phases of populism", including the measures traditionally considered neoliberal:

> **Phase I**: Gross domestic product increased, real wage growth, employment growth, inflation rate controlled and import relieves the scarcity; **Phase II**: bottlenecks by: strong demand growth, little foreign exchange (to pay for imports), inflation Increases, wages unchanged, increased subsidies and increased deficit; **Phase III**: causes, capital flight, demonetization economic (barter), fiscal deficit increase, subsidies decrease, inflation increase, and real wages fall drastically, and **Phase IV**: government starts stabilization, IMF Aid and BM declining real wages, control of inflation rate, break companies, and the capital flight following the violent overthrow of the government.
>
> (p. 103)

Actually, Cortés identifies the flow of crisis/stabilisation with the naturalness of the logic of capital. The macroeconomic measures and their consequences presented as phases show one of the points that I want to make evident in this chapter: populism is an indissoluble part of the logic of capitalism, and neither its reform nor its transformation.

If one reviews the passage from phase IV to V, and especially the last characteristics of V, it can be seen how logic indicates that it is on the verge of a new populist process of crisis reduction.

Another way to observe our affirmation regarding the crisis, and populism as a solution, is to notice the flow and "re-flow" of the political parties considered populist in the last decades of the previous century and the first decades of the present century.

The meaning of the wave of populism in the last 20 years can be understood, as Rodrik argues, according to various factors but it is an unavoidable reality:

> It may seem like populism has come out of nowhere. But the populist backlash has been on the rise for a while, for at least a decade or more .... More importantly, the backlash was perfectly predictable. I will focus in this paper on the economic roots of populism, in particular the role of economic globalization. I do not claim that globalization was the only force at play – nor necessarily even the most important one. Changes in technology, rise of winner-take-all markets, erosion of labour-market protections, and decline of norms restricting pay differentials all have played their part. These developments are not entirely independent from globalization, insofar as they both fostered globalization and were reinforced by it. But neither can they be reduced to it. Nevertheless, economic history and economic theory both give us strong reasons to believe

that advanced stages of globalization are prone to populist backlash. I will examine those reasons below.

(Rodrik 2017: 4)

What can be clearly understand is a growing tendency of the populist parties, and how that it is associated with the increasingly frequent crises of the capitalist system. As Rodrik (2017) says, "The populist backlash may have been a surprise to many, but it really should not have been in light of economic history and economic theory".[1]

From another point of view, Demertizis points out some characteristics about populism and crisis:

> In his attempt to explain the emergence and the chances of the far right European populist parties during the period 1990–2000 (i.e. FPÖ, Ny Demokrati, Republikaner, Front Nationale, Schweizer Volkspartei, Lega Nord, Vlaams Blok, etc.), Hans-Georg Betz uses the concept of resentment. He states (2002, pp. 198–200) that in the early phase of their appearance they were greatly buttressed by the defuse grievances of working-class and lower-middle-class electorate against globalization, the immigrants, the fiscal crisis of the welfare state, politicians' corruption and so on.
>
> (Demertzis, 2006, pp. 114–115)

The connections between capitalism's crisis and forms of populism make us think precisely about neoliberalism. In this vein, this chapter tries to examine how there is a connection between neoliberalism and populism, and that the latter is a "surpassed" modality of capitalist crisis management that the elaborated list gives us the necessary elements to understand.

For many years now, both intellectuals and neoliberal politicians have maintained that populism is the force that produces crises, the causes of institutional and economic breakdowns. In this chapter it is intended to show just the opposite: it is precisely the breaks of neoliberalism that come to compensate for populism, and this has, in the twenty-first century, been transformed into its highest stage.

Current situation of neoliberalism

Over the last 20 years, at least, democracy by consumption has elaborated the stage for neoliberalism as a politics of sensibilities. Maybe the best example is that of the World Bank Report (1997) about "changes on State roles" that announces the "new" relationships between market, state and social policies (Chhibber et al., 1997).

The sixth phase of neoliberalism begins with the "progressive government" and the resignation to capitalism as an unchangeable social system. Beyond all types of discussion, one concrete and "hard core" element of progressive government was the acceptance of their reformist character. In the inner logic of this acceptance lives the deepest consecration of reality (Delgado, 2016).

The market needs state aid to warrant the capacity of people to consume, and by this ensure the all-dominant economic system. One way to characterise this phase is to take into account the economic role of immediate enjoyment through consumption: the mixture and sum of mimetic and compensatory consumption have resulted in a state that warrants market profit.

It is in this way that the components of a political economy of morality that allow the reproduction of the capitalist system today are the consequence of the birth and death of neoliberalism as a political regime. In the same vein, the alluded components make possible the consecration of enjoyment as the centre of life and the forgetting that it is possible to change the social world. Anxiety, non-movement, freedom without autonomy, and a daily life ordered around compensatory and mimetic consumption is the end of neoliberalism history.

In this context it is clear: neoliberalism is part of the populist cycle and flow, but in a deeper sense we need to see some of the characteristics of democracy by consumption, especially the assisted exploitation and subsidised consumer.[2]

Progressive governments of the last decade have configured a new way to "embody" the components of neoliberalism: economic growth is delivered through consumption policies; related to these are applied conditional cash transfers, and a big part of GDP comes from the international price of commodities. The phantoms of poverty are disbanded or diluted through the fantasy of compensatory consumption. As in Foucault's analysis, social policies became a central key for managing class conflict. A paradox has been installed: government that comes from social movements dissolves collective action into the public and private spectacle of enjoyment.

The convergence between the public policies oriented to ensuring corporate profit and the expansion of the market to commodify the sensations have created the main opportunity for the reproduction of neoliberalism's style of capitalism.

The consumers there aren't citizens, at least in the "classical" way. They are enjoyment seekers without political motivation. More or less two decades of progressive government have resulted in the insulation of people in consumption.

The tension between citizen and consumer takes shape as the "*subsiadano*",[3] a key part of the challenge to understand the "political" for the social sciences in the twenty-first century.

The creation of the citizen took place in France and the "West" throughout the long period from the French Revolution to the Second World War. The introduction of the practices/narratives of rights was incorporated into the political economy of the already-established moral positionalities of producer/consumer and customer/citizen who thinkers from Parsons to Habermas intuited as the axes of the post-war state.

Rightly called the successive crises of the welfare state are the provisions of consumer, producer and citizen which undergo a profound transformation.

The worker/producer from Fordism, through Toyotism, reaching to Uber-isation and digital work has been characterised by disappearance, casualisa-tion and permanent transformation.

The consumer has existed for more than half of twentieth-century history: in the form of "comfort seeker", or like the "one-dimensional man" (*sensu* Marcuse), or as a serial addict to fashions, or in its latest form as dependent upon immediate enjoyment. There is a direct passage of the citizen-voter of governments to the quasi-universal elector via consumption and "acceptable" sensibility. In this way state practice implies a transversal orientation to gender, ethnicity and class actions that continue with support programmes directed towards compensation policies.

These tensions between producer/consumer/citizen are evidence of the basic "organisational" features of a state that is undergoing a profound transformation.

By 2015 Latin America had more than 135,000,000 people receiving con-ditional income transfers. To this figure must be added the millions of indivi-duals who are "owners" of other programmes and living in precarious conditions of "assisted liability" by state intervention or omission.

Also, if we include the numbers of citizens who receive subsidised trans-port, energy and/or basic services (just to mention three activities), the mil-lions of subjects increase and multiply, and show a relentlessly structured trans-classed subsidy. Suffice it to recall only the declared intention of the World Bank to show an increase in the middle classes. Finally, is important to incorporate in the analysis the propensity for sustained high participation in public employment in the formal labour force which in some places has become the only source of income.

If we consider the current conditions of state aid and state action strategies, we are not only citizens in the Global South, but we are subsidised. In this context it becomes relevant for democracies to include citizens who are "content" (using a medical analogy) and "happy". From this point of view, it is state practices that establish the resignation (it is not possible to effect any transformation) as a consequence of the logic of patience and waiting (you must to wait your turn) as "civic virtues".

Relations between subsidy and citizenship are established through:

1   *A systemic consecration of rights as untying narratives/with the real indi-vidual (*sensu *Marx).* Both in the "axes" of the central countries and the Global South there has operated an increase of the "consecration of rights" without effective guarantee of their update, and it is possible observe how the "generations of rights" occur, encompassing increasingly aspects once considered to be the realm of the private or something pro-tected or regulated by the state: consider the right to enjoy implants and robotics and/or informational interfaces. And in its (painful) obverse we see hundreds of thousands of victims of multiple wars for which the aforementioned "implants/interfaces" become necessary complementary

forms of humanity. More dimensions of life with codified rights do not imply more resources or dignity in reality.

2   *Redefining the connections between "lack of", consumption and suture.* The *subsiadiano* votes as he buys, buys how he feels and feels like he is assisted. The state is no longer the only actor responsible for public policies; now it is the market that must ensure the subsidised spaces of everyday life. The market as an agent that elaborates the condition of possibility of consumption must guarantee the possible sensibilities of the *subsiadiano*.

The expansion of state actions is proceeding through the market: it is agreements with private capital that determine subsidised areas as hubs for the long term reproduction of capital.

The state "does consume" to make the company responsible for the producer/consumer synthesis, thus creating the condition of possibility of a rational *subsiadiano*: an "opportunity seeker" defined by their cunning to achieve consumption at lower prices and to improve the purchasing power of their income.

The *subsiadano* is a fundamental component of the financialisation of daily life: everything in instalments, all on credit. This represents a whole portion of American middle-class white man who lives his life in instalments and who "can lose everything". Capitalism retraced its steps by re-structuring the relationships between savings/consumption/sacrifice/luxury/credit. The twenty-first century is the dialectical tension between these practices through a political economy of morality whose backbones are sensibilities.

Collecting, receiving and "use" are the political practices of *subsiadano*: collection by a public policy of aid, receiving conditional income transfers and use of the benefits of agreements between the state and the market.

Participation in cooperatives, micro-enterprises and canteens, among other practices, make subjects collect a reward monthly "in order month" (and with a credit card) that installs a systemic instability between employment and work. Updatable subjects receive amounts of money with which they must take action to ensure the continuity of the aforementioned reception: bring children to medical checks, make them attend school, and so on. The *subsiadanos* "enjoy" subsidies given by the state to companies supplying gas, water, energy, public transport, and so on.

Behind this enjoyment there is the systematic increase of corporate profits disguised as aid to citizens.

The assisted exploitation is the superior phase of flex-exploitation and depressed desire. The compensatory and mimetic consumption works as a vehicle of deep and inadvertent sensibility construction. The dispossession of a capacity to make connections between desires, pleasure and enjoyment is the pillar of normalised society.

The perfect milestone to a democracy by consumption and assisted consumption is that of exploitation through aid. The new social policy function is to make possible the extraction and refocussing of bodily energy. These two processes are made reality by immediate enjoyment through consumption: the people lose desire and pleasure and replace them with instantaneous and indeterminate enjoyment.

The logic of global capitalism implied the shift of colonisation from the external/"natural" world to the inner/subjective world: exploitation, expropriation and dispossession are the three faces of neoliberalism as a politics of sensibilities.

In the context of what has been argued so far, it is possible to understand how populism is associated with the crisis of capitalism and how so-called "neoliberalism" comprises no more than renewed ways of acceptance and deepening of democracies by consumption within the framework of societies normalised in the enjoyment immediate through consumption.

In the next sections, two concomitant but different processes are developed, which give rise to a better understanding of the insufficiency that the division between right and left populism implies as an explanatory component of the phenomenon: compensation and lumpen progressivism

Populism and compensation

The "history" of populism is the history of compensations invented/reinvented before the consequences of the unidirectional appropriation of resources – compensation that becomes a democratic objective par excellence.[4]

In the geopolitical spaces of existence of the Global South and Global North, a consumption democracy is currently being applied. Some of the modalities of this democracy imply populism as the guarantor of the state-mercantile consumer.

The system consists of buying a vote; a man is no longer a vote. The *subsiadiano* represents the tension between actor, agent, individual and citizen. Its strengths and weaknesses are reflected in the market guaranteed by the state.

The form of democratic government becomes an instrument of conflictual pacification, the rule of law a guarantor of the unequal distribution of subsidies and the neoliberal and populist government regime implies the dialectic between suture and debt, between bridge and break.

The political regime has been transformed by the way in which a politics of sensibilities ensures the possibility of reproducing the equilibrium and stability of a political economy of morality that solidifies the capitalist structure.

Viewed "from below", "from personal existence", "from infrastructure" and from an intersectional perspective, twenty-first-century populism as the highest stage of neoliberalism implies, at least, three characteristics: a) identify, select, produce and massify the means of enjoyment are the first objectives in terms of the basis of consumer democracy; b) sustain and consolidate the expansion of society 4.0 in three senses: support the digitalisation of the productive system, prepare the modification of work-management strategies and expand control and securitisation mechanisms; and c) the system consists in transforming the purchase/consumption into a substitute for participation/inclusion.

Consumption democracy has expanded in such a way that citizens demand not only compensation in terms of health, education, justice or any other state service, but mainly claim to guarantee the conditions and infrastructure for consumption. In this context citizens vote every time they buy; the old neoliberal theory today is complemented by the slogan that just because a politician is democratic, this does guarantee the production of goods and services to be bought. In this way a virtuous circle is specified; the way in which a state guarantees the successful expansion of the market coincides with the needs of consuming for lay-citizens. In this way, the state takes the responsibility of reproducing capital through reproducing its role as guarantor of consumption. This is how, today, the right is reinvented as a populist left and/or as an upper phase of neoliberalism.

So-called right-wing populism is nothing more than the result of the history of the compensation mechanisms implemented by capitalism in the face of crises. Populism as the highest stage of neoliberalism does not begin in the last decade, but is the result of an internal history in which we can see the displacement of its reformist content to its current guidelines of guarantor of consumption and normalisation. Populism can be understood from the history of the various forms of compensation that it has carried.

If we return to the assumption of the connection between populism and the risk of dissolving the forms of segmentation of structural conflicts, we can visualise the following moments, according to the modifications of the compensations:

a   a stage of mutualist recompositing to the effort of the workers and the poor classes;
b   a period of philanthropic and private benevolence;
c   a state suture phase through social programmes;
d   a stage of inter-class and intra-class compensation through social policies;
e   a phase of compensation through consumption;
f   a compensation moment to capital through the consumer subsidy.

These last two are at the same time facets of the dialectical compensation and subsidisation emphasised in the first two decades of this century.

a   From the point of view of compensation, populism implies the history of the ways of "closing", "sewing", "suturing" the failures, faults and/or injuries caused by the capitalist system. In this direction it can be understood as one of the solutions that capitalism finds, either through the state or civil society, for self-generated problems through exploitation and expropriation. In a first step, mutual societies, mutual aid associations and the cooperative movement appear not only as an alternative to capitalist production and "security" but also as a first form of privatisation of the sutures of market failures. The workers help and contain the workers themselves. Solidarity is thus framed in a dialectic between a collective practice of class self-care in the face of a relief from the ruling class. State regulations allow, accompany and encourage intra-class compensation and self-responsibility. Political groups, workers' associations and groups of benefactors take these compensations as the axis of their demands and practices and thus the remote origin of what we are going to call lumpen progressivism is also established.

b   In one more band of the Möbius strip that implies the connections between populism and compensation philanthropy and beneficence are transformed into new ways of "going to the people". The religious, philosophical and theoretical reasons that philanthropy used were based on "natural disadvantages", on the persistence of the state of barbarism, on the divine will, and proclaimed the need for individual compensation for the failures and damages caused by said "causes". The gaze was placed on individuals beyond their class, ethnicity or sex, education, socialisation in universal values and religious or philosophical training should complete the material efforts to help the poor and abandoned. The market and the state had no responsibility and it was civil society that should help disadvantaged individuals. It is important to note that in this level of compensation appears the figure of professional aid. What we now call social work has its origins in the need to make systematic and transferable compensation practices that the heat of Taylorism imprinted in the organisation of work claimed scientificity.

c   At a level of overcoming, the state appears as a vehicle for compensation, thus creating the institutionalisation of the state's duty to compensate what the "battles" of the market leave as a balance. This phase implies the convergence of the new management techniques, the innumerable faults and wounds produced by the system, and the increase in the discussion of the "social question" by unions, parties and churches. Capitalism gives clear signals: the state must be an instrument to diminish the inevitable damages of the system and improve the incorporation of the citizens into the market as producers/workers. The reproduction of labour over time begins to be an institutional concern. The state apparatus and the bureaucracy are oriented to diminish the conflictual potential through the social programmes of attention and assistance to the poor, hungry and migrants that leaves the system in the working-class neighbourhoods without any attention.

d   In a turn towards an improvement of compensation, the state begins to manage and promote the transfer of resources to all classes in an intra- and inter-class process. The state must guarantee, at least, the stability of profit over time to entrepreneurs subsidising the energy, communications infrastructure and so on, and provide the working classes with education, health and recreation as a suture of the deep deprivations in which they live because of capitalist exploitation. The welfare state is a compensatory agent and the central model of all forms of populism beyond the current transformations. At the intersection between professionalisation and institutionalisation, compensation is transformed into social policies whose explicit objectives are to diminish the conflictual risk, to increase dependence on the state and to serve as an engine of demand in the market. It is at this point where the economic conceptualisation that recommends promoting consumption so that the internal market grows and so that this later impacts on economic growth begins to be part of the central features of state structures.

e   Within the framework of the emergence and consolidation of normalised societies in the immediate enjoyment through consumption, states begin to use consumption as a mechanism of compensation and privatisation of satisfiers. On the one hand, enrolled in the complex web of social plans and programmes that grant and/or facilitate the availability of money or access to informal credits, the state generates and promotes "inclusion by consumption" as a suture of market failures and the abstinence of the state itself. The main intervention of the state in society is commercial: enable the largest number of people to buy. On the other hand, the state promotes the generation of income and other satisfactors not only through a process of privatisation but also to stimulate mass consumption. In the original model of the welfare state, education, health and housing implied the production of satisfiers. It is the market players who make those goods and services that the *subsiadano* demands be guaranteed by the state.

f   In the last phase, but surely not the final one, the state provides the elements and resources to ensure the place of the private in the provision of satisfactions and the increase of the profits of the private capital that lives in and from state resources. In some sense the state happens to be a compensator of loss of profits and guarantor of the stability of the profits of the companies now as a vehicle of massive compensation. The state pays private companies to build and manage roads, airports, public sports facilities and also for drinking water, sewers, streets, energy, and so on. All infrastructures compensate the large companies directly with the financing from everyone's taxes and are the support of the satisfiers of the different social classes that must also be compensated.

It is in the narrative direction that one can understand how the compensation crises caused by restrictive measures to the expense and monetarisation of economic measures are those that "cause" populism, and not the other way around.

It is in the direction indicated that populism through compensation becomes a higher phase of neoliberalism. In this context, to distinguish right or left populism is not important; the new embodiment neoliberalism becomes the populism which consecrates the actualisation of all coloniality: conquer, occupy and expand.

One of the forms that populism acquires as the highest stage of neoliberalism is its transformation into a lumpen progressivism and its consequences for the conformation of a lumpen democracy. It is this very characteristic that prevents talk of left-wing populism and directs a radical critique of its consequences.

Lumpen progressivism (LP) and lumpen democracy (LD)

In the twenty-first century lumpen progressivism is a fundamental part of the practices of the feeling of a populist political regime that in its tension with neoliberalism ensures the reproduction of the *subsiadiano*.

There is a relationship between lumpen progressivism and lumpen democracy as ways of understanding populism as the superior formation of neoliberalism. In this way lumpen progressivism is the political regime that makes possible the differential appropriation of the ideological surplus value. One way to understand the current connections between lumpen progressivism and populism is to take up the connections between the recent history of collective actions, social movements and the history of politics of sensibilities.

Lumpen progressivism has at least three characteristics: a) it reproduces a state of dependence on colonial and Eurocentric interpretations; b) it limits interpretative schemes to empty boxes where all the facts must be imputed; and c) it devalues the content and scope of popular practice.

In short, it is a progressivism subordinated to the material conditions of the development of capitalist ideas: the trivialisation of the notion of all, the scrap of radical transformations and the perversion of rights that have become welfare-oriented policies for consumption.

The progressivism of the twenty-first century brings together the same characteristics that Andre Gunder Frank gave to the lumpen bourgeoisie: a) it depends on the ideological practices accepted and consecrated outside the system in which it operates; b) it bases its power on the sale of commodities; and c) it generates a set of policies of underdevelopment assisting the consumption of the masses:

> The colonial and class structure is the product of the introduction in Latin America of an ultra-exploitative export economy, dependent on the metropolis, which restricted the internal market and created the economic interest of lumpen bourgeoisie (producers and exporters of raw materials). The interests in turn generated a policy of under or lumpen development for the economy as a whole.
>
> (Frank, 1974, p. 14)

The most important consequence of lumpen progressivism is the construction of a lumpen democracy characterised by the creation of the tensions arising between the fantasies of backwardness and the fantasy of progress, the massive use of social assistance as an instrument of conflictual decline and the fictionalisation of the direct contact of the leader with the mass:

> Perhaps the most striking event is the habituation to atrocities and calamities, the exhaustion of popular struggles and the rise of what might be called lumpenradicalism, that is to say a form of nihilism that goes for radicalism. Lumpenradicalism, whose rise is favored by access to digital

technologies, operates by annexing the categories and languages of emancipation and their diversion into causes and practices that have nothing to do with the quest for freedom and freedom. equality or the general project of autonomy.

(Mbembe, 2017)

Populism as lumpen progressivism will never be a project of personal autonomy and communitarianism; it is nothing more than the banalisation of what there is in it as a transformer.

There are many features to characterise the LP, but in this space I want to emphasise three: a) the acceptance of capitalism as the unique regimen of accumulation and power; b) the articulation with the international commodities market; and c) the non-recognition of Enlightenment and vanguardism.

a) The acceptance of capitalism as the unique regimen of accumulation and power

The meaning of acceptance is defined through three axes: LP never discusses the elimination of the causes of inequality, it doesn't have another proposal for the management of property and resources, and has in the process of struggle a complex mechanism of competence and meritocracy. Acceptance is deep, or superficially senses practices that involve the incapacity of persons to change the world. Acceptance is the naturalised sensation of an impossibility for structural change from institutions and social organisations.

In this context LP refuses to take into account the possibility of a radical transformation of the causes of inequality and oppression. In the vein of neo-colonial religion, LP is a sophisticated form of resignation. The bottom is the negation of deep change, the top the fantasy of the human face of capitalism. The problem is not to erase the difference between the richest and poorest people; it is the "distribution" of this difference. The LP distributive model is a "sensible" means for accepting reality. Nor are either property or possessions placed in question; LP's proposal is to assist injustice by means of state resources that come from taxes that are paid by the poorest people. LP's narrative consecrates the notion of "redistribution" at the same time as a naturalised perception that there are some persons who have much more than others. The causes of inequality, expropriation and dispossession are never the object of public policies.

Acceptance implies the negation of alternatives, a refusal of other ways of doing things, the impossibility of different paths and patterns.

The management of property and resources is not placed in a changing horizon by LP; the compensatory model is built on the acceptance of labour exploitation. The narrative is about the consensus about size and weight of compensation. LP's contemporary big question is: how is it possible to move from a well-being to a well-feeling state? Under human rights discourse, LP shapes the propositional model of acceptance of injustice and inequality in the system of property and possessions. In LP's horizon there is no place for a change in the world of work, or the modification of a politics of sensibilities that is based on having a unidirectional pattern of accumulation.

Acceptance is a type of socialisation grounded on never questioning the taken-for-granted of social reality, is a way to adapt personal desire to social acceptance and expectations.

LP shapes as a revolutionary narrative the acceptance of competence, meritocracy and destination. Fight is the consecrated obverse of a Darwinian model of savage capitalism. The person on and for the "battle" is an archetype of the producing/consuming individual. The other is the abject, is the potential enemy, is the radically different. From an intersectional viewpoint, the other is dangerous and the LP narrative tries to "forget/hide" this feeling-practice behind the "discourse of rights" fantasy. LP's epic politics of sensibilities cannot see the person's emotions and their existential material conditions.

b) The articulation with the international commodities market

The process of social structuring on a planetary scale as I have just described it involves other processes that intersect: the financialisation of the lives of billions of people both in the Global North and the Global South, the consolidation of the so-called revolution 4.0 and the dependence on "emerging markets" for the sale of commodities.

In this framework the LP has constituted an LD in and through consumption. These three characteristics of the macroeconomic model of consumer democracy converge, among other factors, in the following: volatility. The bancarisation of lay-people through electronic money, the creation of reticular and "asphyxiating" credit structures, the creation of financial investment centres and tax havens, and the extension of their use are connected by dependence on the uncertainty of the international financial system. The hundreds of thousands of new or even non-existent jobs of the revolution 4.0, together with its virtual and techno-dependent nature, make uncertainty one of its constituent features. The commodities in all their versions are characterised by being susceptible to the variation of their international price, to the geopolitical contexts of predation and the technological variations of their production and management. From gold, through soybeans to lithium, they imply a high degree of variability over time.

c) How to be postcolonial but still be a believer in the Enlightenment

There are three axes that make lumpen progressivism an excellent ideological practice of populism today: 1) the paradox of trust and dependence of the leader and the claim of the fantasy of rights that consecrate autonomy; 2) the production of a politics of sensibilities that marks a convergence of collective emotions and practices in contradiction of the promotion of the scale of immediate enjoyment through consumption; and 3) the contradictory conception of the capacity of the mass media and social networks to act on the people supported in an idea of a "new vanguard" that is capable of realising when those media and networks "deceive/manipulate people".

1   Beyond the individual sympathies and possible positive evaluations of many progressives (except for Uruguay) there is an incontestable reality: neither Evo Morales, nor Cristina de Kirchner, nor Lula da Silva (just to mention three of the most representative figures) have been able to create new leadership based on the democratisation of mass movements. The internal divisions of their parties, the results of elections and the high rate of distrust of politics in all Latin America are some observational elements that guarantee this claim.

    In the same direction and in connection with the view on the media and networks, direct contact with people was privileged as a vehicle of interaction with citizens. The mere fact that they began to use their first names when dealing with and from the leader, and the permanent appearance of the same in the networks, provide an indicator of the use of an old strategy of "proximity" based on emotionality, condemnation and the sacralisation of the leader's skills.

    The autonomy proclaimed by the official narratives and by the democratically approved laws have been dissolved in the indispensability of the leader to guarantee these rights. He or she is the only one to guarantee the continuity of the progressive process.

2   In the LP processes, the street manifestation is promoted, as well as participation in public spectacles and mobilisation for national days. Paradoxically with this (and as we have explained above), consumption and associated enjoyments are promoted and rewarded. It is the LP governments themselves that strive to "create new consumers" and the consumer generates a structure of sensibility associated with individualism and isolation.

    Emotions as constitutive practices of the policies of collective manifestations become a trap: on the one hand they imply the escape route of the accumulated tensions individually in work and daily life, and on the other hand increase the dependency upon governmental actions for management of sensibilities.

    From this tension, it is possible to observe that there emerges and grows non-participation, massification of solidarity, neo-humanisms and distrust in the public. The subjects become an auditorium of shows where being with others is governed by the rules of "live-the-spectacle".

3   The LP as narration of a self-fulfilling prophecy believes and teaches that the media manipulate people and that is why these agents do not act to support the "transformative" initiatives of the LP. The explanation is simple: the media concentrated in the hands of a few owners produce false news, fictions and rumours that deform the perception of the people who receive the message. This deformation of reality operates like a

bandage that prevents people from seeing the true reality of transformations of the government and its leader.

It is in this context that the LP auto-postulates itself as the only avant-
garde capable of identifying the falseness in the media, explaining to the
people who are really living and taking measures against the "hegemonic" media. In the name of the "aggression" of the media, it is self-
justifying that a "popular movement" is self-described as the only one
capable of defending individuals by illuminating and explaining the real.

Conclusion

The compensation of the systemic failures operated by populisms as the
highest stage of neoliberalism leads to the massive expansion of normalised
societies in enjoyment through consumption. The constitution of lumpen
democracies through lumpen progressivism implies projects of dependence
and coloniality that have become politics of sensibilities. The superposition of
democracies by consumption assisted exploitation and subsidised citizenship
with the compensation and trivialisation of progressivism constitute an iron
cage for people and collective practices. Inclusion with consumption generates
a devalued form of democracy where the central issue is the creation of a set
of experiences that guarantee the connection of purchasing a vote.

The triumph of neoliberalism as a politics of sensibilities implies the passive
acceptance that reality cannot be transformed, and that what is pragmatically
relevant is enjoyment. In the consequences of spectacle, sacrifice and enjoyment of contemporary populism, it allows the massive acceptance of the
unmodifiable character of the social system.

Within the framework of what has been analysed throughout the chapter, it
is important to note that the history of the connection between populism and
compensation involves a history of the revolution of expectations. These
expectations are the way that, since the beginning of the last century, capitalism has elaborated and managed to "seduce" classes and class fractions that
were not part of the distribution of resources. From the revolt of aspirations,
through aspirational consumption, to the revolution of expectations, the
management of sensibilities has become one of the axes of populism today.

In this vein, populism as the management of emotions implies the massive
and planetary expansion of a special political economy of morality and the
globalisation of a racialised and colonial form of sensibilities. This is a coloniality of emotion that comes from global populism as a superior stage of
neoliberalism, in the sense that Quijano (2000a, 2000b) suggests about the
coloniality of knowledge. It is a new form of coloniality of power.

This theoretical, historical and epistemic view of populism would not be
complete if at least we did not affirm that this "victory" of neoliberalism
under the mask of populism as its highest stage, and the politics of the sensibilities associated with it, has its systematic denial and break in the interstitial

practices that millions of subjects perform every day, especially the one that comes from love as collective action (Scribano, 2017).

Notes

1 https://voxeu.org/article/economics-populist-backlash
2 For growing the global scale of social protection see the CFR World Social Protection Report (WSPR) 2017.
3 Combination in Spanish of subsidised and citizen.
4 As is obvious here, a synthesis of the aforementioned history is made, a task that has been inspired by various texts, but I want to make particular mention of the works of Federici (2004), Donzelot (1984) and Titmuss (1975).

References

Chhibber, A., Commander, S., Evans, A., Fuhr, L., Kane, T., Leechor, C., Levy, B., Pradhan, S. and Weder, B. (1997) 'World Development Report 1997: The State in a Changing World', *World Development Report*. Washington, DC: World Bank Group.

Cortés, R. O. (2010) 'The Danger of Populism Macroeconomic Populism in Latin America, is Colombia the Exception?', *Prolegómenos. Derechos y Valores*, 23(26), July–December, pp. 99–122

de La Torre, C. (2017) 'Populism in Latin America', in C. R. Kaltwasser, P. Taggart, P. Ochoa Espejo and P. Ostiguy (eds) *The Oxford Handbook of Populism*. Oxford: Oxford University Press. doi:10.1093/oxfordhb/9780198803560.013.8

Delgado, J. O. (2016) 'Sociedades postneoliberales en América Latina y persistencia del extractivismo', *Economía Informa*, 396, pp. 84–95.

Demertzis, N. (2006) 'Emotions and Populism', in P. Hoggett, S. Clarke and S. Thompson (eds) *Emotion, Politics and Society* (pp. 103–122). Basingstoke: Palgrave Macmillan.

Donzelot, J. (1984) *L'Invention du social: essai sur le déclin des passions politiques*. Paris: Fayard.

Dornbusch, R. and Edwards, S. (1991) *The Macroeconomics of Populism in Latin America*. Chicago: University of Chicago Press.

Federici, S. (2004) *Caliban and the Witch: Women, the Body and Primitive Accumulation*. New York: Autonomedia.

Frank, A. G. (1974) *Lumpenbourgeoisie: Lumpendevelopment*. New York: Monthly Review Press.

Jansen, R. S. (2011) 'Populist Mobilization: A New Theoretical Approach to Populism', *Sociological Theory*, 29(2), June, pp. 75–95

Mbembe, A. (2017) 'Le lumpen-radicalisme et autres maladies de la tyrannie', *Le Monde*, 28 December. Online at: https://www.lemonde.fr/afrique/article/2017/12/28/le-lumpen-radicalisme-et-autres-maladies-de-la-tyrannie_5235406_3212.html (Accessed 3 March 2019)

McKnight, D. (2018) *Populism Now! The Case for Progressive Populism*. Sydney: NewSouth.

Mudde, C. and Kaltwasser, C. R. (2017) *Populism: A Very Short Introduction*. Oxford University Press.

Quijano, A. (2000a) 'Coloniality of Power, Eurocentrism and Latin America', *Nepantla*, 1(3), pp. 533–580.

Quijano, A. (2000b) 'Colonialidad del poder y clasificación social', *Journal of World System Research*, 1(3), pp. 342–388. Rodrik, D. (2017) *Populism and the Economics of Globalization.* London: Centre for Economic Policy Research. Online at: https://cepr.org/active/publications/discussion_papers/dp.php?dpno=12119 (Accessed 28 April 2019)

Savarino, F. (2006) 'Populismo: perspectivas europeas y latinoamericanas Espiral', *Estudios sobre Estado y Sociedad*, 13(37), September–December, pp. 77–94.

Scribano, A. (2017) *Normalization, Enjoyment and Bodies/Emotions: Argentine Sensibilities.* New York: Nova Science Publishers.

Scribano, A. (2018) *Politics and Emotions.* Houston: Studium Press Llc.

Scribano, A. and Lisdero, P. (eds) (2019) *Digital Labour, Society and Politics of Sensibilities*Basingstoke: Palgrave Macmillan.

Scribano, A., Timmermann López, F. and Korstanje, M. E. (eds) (2018) *Neoliberalism in Multi-Disciplinary Perspective.* New York: Palgrave Macmillan.

Thirkell-White, B. (2009) 'Dealing with the Banks: Populism and the Public Interest in the Global Financial Crisis', *International Affairs*, 85(4), July, pp. 689–711.

Titmuss, R. (1975) *Social Policy: An Introduction.* London: Pantheon Books.

Index

Printed in the United States
by Baker & Taylor Publisher Services